Lecture Notes in Chemistry

Edited by G. Berthier M. J. S. Dewar H. Fischer
K. Fukui G. G. Hall J. Hinze H. H. Jaffé J. Jortner
W. Kutzelnigg K. Ruedenberg

37

Kjeld Rasmussen

Potential Energy Functions in Conformational Analysis

Springer-Verlag
Berlin Heidelberg New York Tokyo 1985

72383227

Author

Kjeld Rasmussen
Chemistry Department, Technical University of Denmark
DK-2800 Lyngby

sep / ov
CHEM

ISBN 3-540-13906-0 Springer-Verlag Berlin Heidelberg New York Tokyo
ISBN 0-387-13906-0 Springer-Verlag New York Heidelberg Berlin Tokyo

Printing and binding: Beltz Offsetdruck, Hemsbach/Bergstr.
2152/3140-543210

SD
7/2/85
AM

Foreword

I get by with a little help from my friends

The Beatles: Sgt. Pepper

This book should have been in Danish. Any decent person must be able to express himself in his mother's tongue, also when expounding scientific ideas and results.

Had I stuck to this ideal, the book would have been read by very few people, and, indeed, appreciated by even fewer. Having it published in English gives me a chance to fulfill one ambition: to be read and judged by the international scientific community.

Another reason is that the majority of my professional friends are regrettably unread in Danish, just as I am in Hebrew, Finnish and even Italian. I want to deprive them of the most obvious excuse for not reading my opus.

Like a man I admired, I will first of all thank my wife. In his autobiography, Meir Weisgal, then President of the Weizmann Institute of Science, wrote about his wife: "In addition to her natural endowments - which are considerable - she was a more than competent part-time secretary." He wrote on, and so shall I.

The book has been edited by my wife. So if the reader finds the layout pleasant as, in actual fact, I myself do, Birgit is to be praised. If there are blemishes, I am to be blamed for not having caught them.

After having thanked the immediate victim of my decade-long preoccupation with CFF methods, who for long periods put aside her own interests to assist me, I shall make the other acknowledgements more explicitly.

Three men meant a lot to me in the formative years of my scientific career, and still do: Niels Hofman-Bang, Flemming Woldbye and Shneior Lifson. Professor Hofman-Bang was Director of Chemistry Department A for many years, and responsible for employing me in

1967. He took a chance then, as I was older than most, and I fear there have been times when he regretted his choice. Returning to academic life after seven years in industry was not dead easy; Hofman-Bang helped and corrected me when necessary.

Professor Flemming Woldbye supervised my Ph. D. (lic. techn.) work 1967 - 1970. During this period he got both his D. Sc. (dr. phil.) and his chair, and it would happen that we did not meet for months. This situation made me admire his capacity for absorbing information, and for making quick and, in the long run, correct decisions. It was he who established the contact between Lifson and me, which was to become so fruitful in several respects.

Professor Shneior Lifson was founder and for many years director of Chemical Physics Department at the Weizmann Institute of Science in Israel. He introduced me unreservedly to the methods for which he, with Bixon, Warshel and Levitt, laid the foundations, and during my stays in Rehovot I always had the benefit of his help and advice. Even more than that our family always felt a warm welcome with Shneior Lifson, his wife Hanna, and the whole department.

Two former graduate students and present friends were indispensable in earlier phases of the work: Svetozar R. Niketić and Steen G. Melberg; one is so in the current phase: Lars-Olof Pietilä.

Niketić came to us for his Ph. D. (lic. techn.) work at a crucial time and was instrumental in the development of chemical formula treatment and of minimisation. He also developed our first potential energy function for coordination complexes and performed a large part of the calculations.

Melberg, during his M. Sc. (cand. polyt.) and Ph. D. (lic. techn.) work, laid the foundations for our present and future work on saccharides. He prepared very conscientiously a survey of the literature on Lipid A and began the calculations with the development of a potential energy function for glucose. He then performed all calculations on disaccharides prior to 1979.

Pietilä is doing a large part of his work for his Ph. D. (dr. philos) at the University of Helsinki with me. He has done most of the extension to the crystal version.

Other students have given valuable assistance since 1973: Poul Bach, Lars Christensen, Niels Hald, Jens Hansen, Oliver Jacobsen, Klavs Kildeby, Jan Larsen. The methods and techniques they introduced are all in use today. This also applies to Dr. Cornelis J. M. Huige in Utrecht who found and corrected a crucial error in the treatment of planar molecules.

Many colleagues and friends at foreign universities and other learned institutions have on many occasions given me the benefit of their advice and criticism when they and their students endured a seminar or a series of lectures. For this assistance I feel very grateful to my friends in Freiburg, Fribourg, Helsinki, Jerusalem, Novara, Oslo, Rehovot, Tel Aviv, Tromsø, as well as Copenhagen and Aarhus.

Among those colleagues I would like to mention three who have contributed substantially to this work.

Professor Camillo Tosi of Istituto Guido Donegani in Novara has installed and further developed the UNIVAC version of the CFF program. He is responsible for the ab initio calculations behind some of the potential energy function developments. Together we work on conformations of amines and of nucleotides.

Dr. Tom Sundius of the Accelerator Laboratory at University of Helsinki made, with Mr. Ernst Kruckow of RECKU in Copenhagen, the UNIVAC version of the CFF. He then modified his program MOLVIB, one of the best programs available for more traditional normal coordinate analysis, so that it can work in unison with the CFF program.

I have always had good assistance at the computer centres of the Technical University of Denmark, the Weizmann Institute of Science, Istituto Donegani, and the Universities of Copenhagen and Fribourg.

It remains to mention who paid for all of this.

The Danish Natural Science Research Council paid the salary of Niketić for a total of 36 months and the computational costs during the years 1970 to 1980.

Two funds have been particularly helpful.

Otto Mønsteds Fond paid for one of my screen terminals and for my colour raster plotter.

The Technical University Fund for Technical Chemistry supported Birgit Rasmussen for about half of the time she worked on these projects.

Three other private funds supported her: Frants Allings Legat, G. A. Hagemanns Mindefond and Holger Rabitz´ Legat.

In addition, substantial support was received from the following institutions: European Molecular Biology Organization, Kulturfonden for Danmark og Finland, Nordic Council, Tribute to the Danes through Scholarship in Israel, Weizmann Institute of Science.

On several occasions, such help, even with small allocations, has saved an ongoing project, when University money was not available.

Lyngby, in July, 1984

Kjeld Rasmussen

Contents

Tables

Figures

1 Introduction

> For unto whomsoever much is given, of him shall
> be much required: and to whom men have committed
> much, of him they will ask the more.

Luke XII, 48

As organiser of one of the many groups that develop methods for conformational analysis by minimisation of conformational energy and related techniques, I have time and again had to answer the following statement. "It´s wonderful to have those magnificent programs; just give me energy parameters for my molecules." It comes from the budding student we try to seduce into our group as well as from the experienced researcher looking for a framework on which to hang his spectroscopic or kinetic observations or for a rationale telling her which of several possible synthetic routes may be profitable.

The usual answer is: "Try any of the well-documented force fields available, and do not change it, if you are reasonably satisfied. If you are not, then develop your own, and be prepared to invest a lot of work."

Such answer is clearly not satisfactory, as many force fields are available for any one class of compounds, and all have been published by authors who in good faith claim to have produced a set of parameters of good reliability with a high predictive capability.

Anyone who has tried to use one or another of the available force fields on a new task knows that very likely it is not sufficient for the specific problem being studied, and also that in certain cases energy formulae, parameters and their units are not consistent.

In order to rectify this situation, to avoid part of an increasing volume of correspondence, and to convey, if possible, some of our own experiences, I have undertaken to write this monography. I have tried to get through the entire field, but I realise that it is so vast that I can definitely not have reached completeness. To those authors whose work, though relevant to my purpose, unjustly and unintentionally has been left out, I offer my apologies. I shall appreciate if readers will advise me of omissions, and not only of the more blatant ones.

The title of the book stresses the application to conformational analysis. This means that force fields developed exclusively for analysis of vibrational spectra are not mentioned in this capacity, but only if they, wholly or in part, have been used in conformational analysis.

In this context, the term conformational analysis shall mean the calculation of equilibrium conformations by minimisation of molecular potential energy using simple analytic functions with associated parameters. Often, though by no means always, such calculations are followed by vibrational analysis and by calculation of thermodynamic and other properties. Thermodynamic calculations may give very good results, but vibrational analysis can never compete with traditional normal coordinate analysis. It seems to be the ambition of vibrational analysts to reproduce, to within a few cm^{-1}, all observed infrared and Raman bands, using all available data for isotopomers. Often they succeed, at the cost of a large amount of cross-term parameters which tends to limit the usefulness of the resultant force field to one compound or at best to a few closely related compounds. Such accuracy is usually not sought by the conformational analyst who is satisfied with a good reproduction of known structures promising reliable prediction of unknown ones, a reasonable representation of the main features of vibrational spectra, and a match of thermodynamic and crystal properties equal to the fit of structures of isolated molecules.

Some years ago, when the idea of this review began to take shape, I nourished the hope of digesting the entire literature and distilling off a parameter set for general use, grouped into terms of bond stretching, non-bonded interactions, etc. Our own experiences forced me to abandon this approach, and I shall stick to the advice given in the second paragraph of this introduction, organising the review according to classes of compounds and to different approaches to modelling of molecular potential energy.

I shall not refrain from critical remarks or even corrections to work published also by other authors when I feel that such comments may be helpful to potential users.

With this statement I stress my wish to be helpful to fellow scientists whose research into the nature and properties of matter may be furthered by resorting to computational techniques that are very simple in principle, yet complicated and labour-consuming in development and implementation.

I want in particular to be of assistance to those colleagues who have shown faith in our work to the extent of actually installing our products and investing labour in making the programs run. In giving them of my best, yet realising that it was not enough, I feel in a way that I owe them more. I provided them with tools for building a structure. Here come the beams, the joists, the planks, and the paper to cover the imperfections. At least, such is my intention.

This book is to some extent a continuation of one by Niketić and myself [190].

In fact, it is more or less what chapter 8 of the previous book should have been. In addition, important new additions to and changes in the programs are described. An example of this is the documentation of the optimisation routines which were to a large extent reformulated in 1981. This topic, so basic to the CFF concepts, is given detailed attention in chapter 7. Another example is the extension to crystals, which also includes optimisation; see chapter 11.

As the book is a sequel to the former in more than one way, references to it shall be in abundance, and I shall take it for granted that the reader has ready access to ref. 190.

2   Nomenclature

What´s in a name ?
That which we call a rose,
By any other name
would smell as sweet.

Romeo and Juliet: Act II, Scene II

2.1   Crucial expressions

Much misunderstanding as to the perception of a text arises from improper definition of the more important terms used. I give here the meanings of what I hope is a full list of those expressions which may not be obvious or which I use in a way slightly different from common practice.

2.1.1   Constitution

Molecular constitution tells us which atoms in a molecule are linked by conventional chemical bonds, see Figure 2-1. The constitution says nothing about the layout of the atoms; the information is entirely topological.

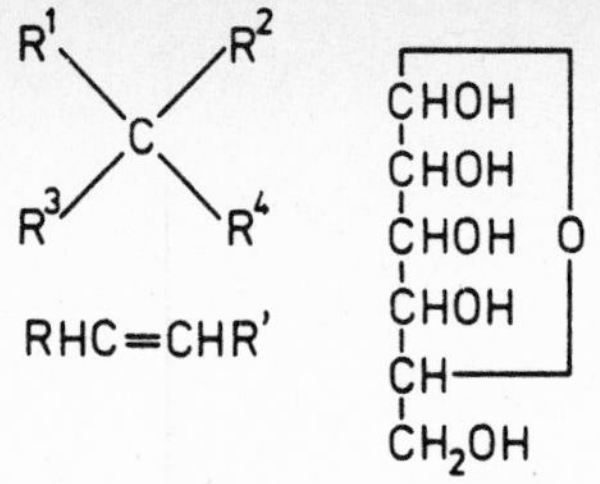

$$R^1 \diagdown \diagup R^2$$
$$C$$
$$R^3 \diagup \diagdown R^4$$

$$RHC{=}CHR'$$

$$\begin{array}{l} CHOH \\ CHOH \\ CHOH \quad O \\ CHOH \\ CH \\ CH_2OH \end{array}$$

Figure 2-1: Molecular constitution

2.1.2 Configuration

Molecular configuration tells us in which of several possible ways the atoms are situated in space relative to each other, when these different arrangements correspond to isomers which can be isolated, see Figure 2-2. Note that you can change from one configuration to an isomeric one only by the breaking and making of a chemical bond, including part of a double bond.

2.1.3 Conformation

This has become a most elusive term. In this book, molecular conformation shall be synonymous with molecular geometry. It is a description of where the atoms of the molecule are situated in space relative to each other, and it may be given as a set of coordinates, internal or cartesian.

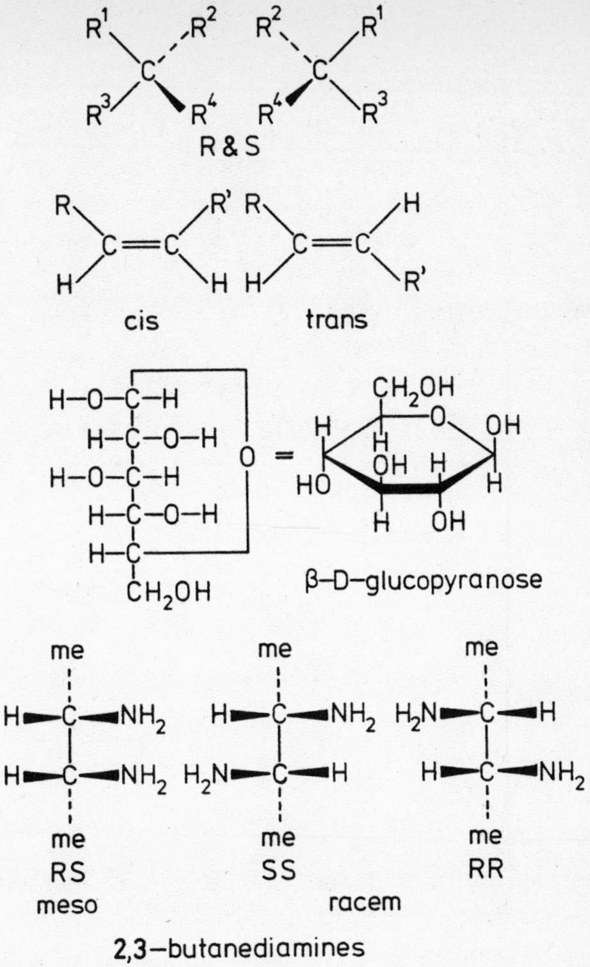

Figure 2-2: Molecular configuration

2.1.4   Conformer

Molecular conformers are, or anyway should be, conformations which can exist, in the sense that they can be observed, at least in principle. Strictly speaking, a conformer must correspond to a minimum in the free enthalpy, and must of necessity be an average over a range of molecular geometry, even at 0 K.

In this book the common practice shall be used: a conformer is a minimum on the potential energy surface. It is therefore a point in conformational space, not a vibrational average. Therefore a conformer cannot, in principle, be observed, but can be determined from observed data.

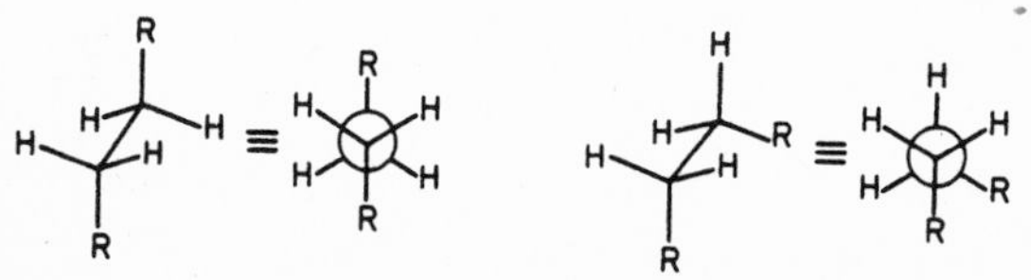

Figure 2-3: Molecular conformation

In Figure 2-3 the anti and one of the two gauche conformers of a substituted ethane are pictured, using the two most practical projections for two-dimensional visualisation.

In both, the C-H and C-R bonds are staggered. The eclipsed conformation is not a conformer, as it cannot exist, so the boat-form of cyclohexane is not a conformer either. One of the gauche conformers of each of the three isomers of 2,3-butanediamine is given in Figure 2-4.

It can be seen that one conformer can be changed to another by rotation around a bond, without breaking any bonds. Even for simple molecules, this may give rise to very large numbers of conformers.

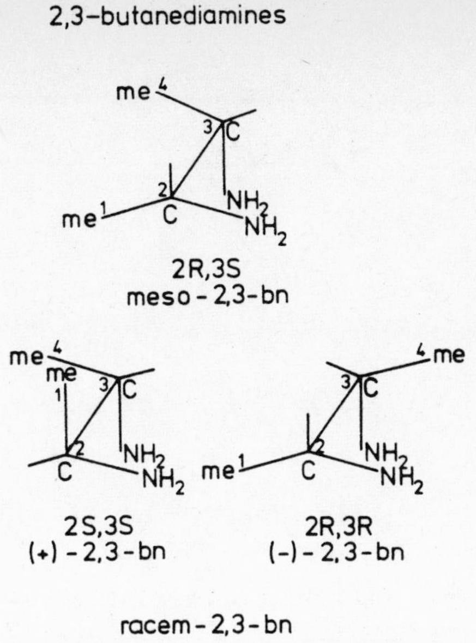

2,3-butanediamines

me ^{4}

^{3}C

^{2}

me ^{1} C NH$_2$ NH$_2$

2R,3S
meso-2,3-bn

me ^{4}
me ^{1} ^{3}C NH$_2$ NH$_2$
C

^{4}me
^{3}C NH$_2$ NH$_2$
me ^{1} C

2S,3S
(+)-2,3-bn

2R,3R
(-)-2,3-bn

racem-2,3-bn

Figure 2-4: Gauche conformers of 2,3-butanediamines

Figure 2-5 shows one of the 729 possible conformers of β-D-glucopyranose.

As a rule, we cannot isolate one single conformer under ordinary conditions. It may be done under special conditions, such as ortho-substituted biphenyls, where steric hindrance may effectively restrict the almost free rotation around the central C-C bond, or at low temperature.

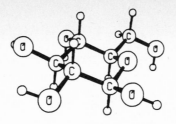

Figure 2-5: One conformer of β-D-glucose

2.1.5   Conformational analysis

A multitude of methods, experimental and theoretical, are used to identify conformers, determine their mutual energy differences, study their geometries, measure their relative occurrences at equilibrium, and find conformational paths of interconversion between them.

Conformer symmetry and geometry, equilibrium ratios of conformers, barrier heights and conversion rates can be analysed by many techniques, mainly diffraction and spectroscopic experiments. Examples are given in a short review by Woldbye and Rasmussen[285].

2.2   Structure

The term "molecular structure" is ambiguous and therefore somewhat elusive. In this book it is used in two senses: in a general sense meaning "one of the well-defined structure types, but an unspecified one"; and in a specific sense meaning "one particular of the structure types". A molecular structure is thus a specification of the geometric arrangement of the atoms of the molecule which has been derived from measurement in some well-defined way.

Various structure types are described shortly in Table 2-1, which is taken, with few changes, from a paper by Kuchitsu and Cyvin [141].

This paper gives clear definitions and descriptions of structures and should be consulted by all. Too many people speak -and write- too often about "structure" without knowing what a structure is and what is in a structure. A discussion by Kuchitsu [142] is illuminating on the differences between different structures.

Table 2-1: Molecular structures

r_o Distances between nuclear positions derived from rotational constants A_o, B_o, C_o for the ground vibrational state.

r_e Distances between nuclear positions derived from rotational constants A_e, B_e, C_e for the hypothetical vibrationless state; A_e etc. being derived from rotational constants for at least two vibrational states, one of which is always the ground state.

r_v Distances between average nuclear positions of vibrational state v.

r_z r_v for the vibrational ground state.

r_s Substitution structure for a set of isotopomers gotten with Kraitchman's formulae. May or may not be close to r_e. The most accurate structures one can find.

r_a Argument in ED scattering intensity, derived from scattering vectors; thermal average values of internuclear distances evaluated with a weight factor r^{-1} in the averaging.

r_g Thermal average values of internuclear distances.

r_α Distances between average nuclear positions at thermal equilibrium.

r_α^o Distances between average nuclear positions at thermal equilibrium in the vibrational ground state. Equal to r_z.

Differences between various structures of propane are exemplified in Table 2-2; and average structures of methane and deuteromethane are illustrated in Figure 2-6.

Table 2-2: Structures of propane

	C-C	C-H
r_a	1.5307	1.1017 (aver.)
r_g	1.5323	1.1073 (aver.)
$r_z = r_\alpha^o$	1.5307	1.0956 (CH_3)
		1.0941 (CH_2)
r_s	1.526	1.091 (CH_3)
		1.096 (CH_2)

2.3   Potential energy function

The meaning of this term is probably evident. As an illustration, Figure 2-7 shows a selection of contributions to molecular potential energy. The list is by no means exhaustive, neither as to what has been used and described in the literature, nor as to what can be handled with the program used in most of the studies described in this book.

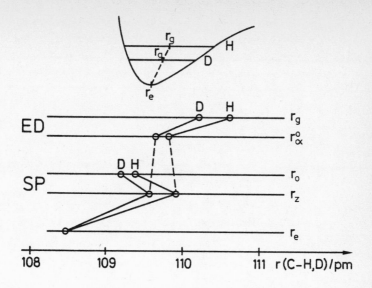

Figure 2-6: Structures of methane

2.3.1 Parameters

This word shall be synonymous with "potential energy function parameters", and its meaning shall be those quantities of a potential energy function which are unchanged from one molecule to another within the same series of minimisation or within one iteration in optimisation.

Parameters are not force constants, and only some of them have the same unit as a force constant. $b_o(C-H)$ is not the equilibrium bond length of any C-H bond in any molecule, and $K_\theta(C-C-H)$ is not the force constant of any C-C-H deformation.

$$V = \begin{cases} 1/2 \sum K_b(b-b_o)^2 \\ \sum D(\exp(-2\alpha\,(b-b_o))-2\exp(-\alpha\,(b-b_o))) \\ \sum (A/b+B/b^5+C/b^9) \end{cases} \text{summed over all bonds}$$

$$+ \; 1/2 \sum K_\theta(\theta-\theta_o)^2 \qquad \text{summed over all valence angles}$$

$$+ \sum (1/2F(d-d_o)^2+F'(d-d_o)) \qquad \text{summed over all 1,3-interactions}$$

$$+ \begin{cases} 1/2 \sum_{n=1}^{N} K_{\phi n}(1+\cos k\phi_1)/N \\ \\ 1/2 \sum K_\phi(1+\cos k\phi) \end{cases} \text{summed over all torsions}$$

$$+ \begin{cases} \sum_{i>j} (A/r^9-Br^6+e_1e_j/r) \\ \sum_{i>j} (Ae^{-Br}-Cr^6+e_1e_j/r) \end{cases} \text{summed over all non-bonded distances}$$

Figure 2-7: Molecular potential energy functions

2.3.2   Potential energy surface

This is a surface in 3N+1 dimensions where N is the number of atoms in the molecule; the extra dimension over the cartesian con-formational space is the potential energy. It may also be a section of the full potential energy surface where most coordinates are kept constant, or even an approximation to such a section. An illustration is seen in a paper [192] by Niketić and Rasmussen. Here potential energy surface for ring flipping in a tris-diamine coordination complex is adequately approximated by a surface in three dimensions and, after introduction of a special reaction coordinate, even as a planar curve.

2.3.3   Force field

A force field is a set of force constants used in the calculation of a vibrational spectrum. In the consistent force field context, a force field is derived from a potential energy function, for a particular conformer of a molecule. The force field will be different for another conformer of the same molecule, and for another molecule. See also ref. 190, section 6.1.

This is in striking contrast to the usual spectroscopic concept of a force field, in which the force constants are - or at least should be - unchanged from one molecule in a set to another.

Some aspects of the differences between a force field in the consistent force field context and as used in conventional normal coordinate analysis are discussed in a paper by Sundius and Rasmussen [255].

2.3.4   Consistent force field

A consistent force field is conceptually rather different from a force field. It is a potential energy function whose parameters have been fitted, preferably optimised in some objective way, to reproduce structures of a set of molecules and, through the force field derived for each molecule of the set, molecular spectra. Other properties such as dipole moment, thermodynamic functions and crystal properties may also be included in the fitting process.

A more philosophical statement of what a consistent force field is may be found in ref. 190, section 1.1, where also the history of the development of the consistent force field by Lifson and his coworkers is summarised.

2.4   Molecular mechanics

This term has become so widespread throughout chemistry, particularly among organic chemists with less than thorough knowledge of this field or of general mechanics that it is probably futile to try to give it a proper meaning. I shall nevertheless try.

Molecular mechanics logically encompasses molecular statics, kinematics and dynamics. It is therefore proposed to restrict the use of the term to situations where it is the intention to denote the entire region of mechanics. Molecular statics is the appropriate expression to substitute for the illogical use of molecular mechanics.

A system which deserves the designation molecular mechanics is described in section 3.5.

3   Potential energy functions: A review

> This we learned from famous men,
> Knowing not its uses,
> When they showed, in daily work,
> Man must finish off his work-
> Right or wrong, his daily work-
> And without excuses.

> Kipling: Stalky & Co.

Since the beginning of conformational analysis with molecular potential energy functions as the main tool a quarter of a century ago, numerous papers have appeared, not only on applications, but also dealing with development of potential energy functions.

From the latest decade we have many reviews, both of the area as a whole and of specific fields. Although I myself feel the need for a comparative, all-encompassing and critical review, this is not my goal here; such a task is possible only for a scientist with deep insight and broad experience on early pension. I shall set myself a less ambitious goal: To indicate the main trends and methods, and mention some notable works from some of the leading groups. Work of our own group is reviewed in other chapters, with one exception (section 3.18).

3.1   Scope

Colleagues have sought my opinion or help, have questioned me and criticised me, on my use of potential energy functions. To generalise a bit, they fall into four categories.

One consists of mainly organic chemists with an outlook somewhat beyond that of elegant synthetic and NMR work. After an introduction to consistent force field concepts and methods, they may ask: "Why don't you just use Allinger's program?" or, "Can't you incorporate Allinger's potentials?" Usually, these people do not mention the other widely known initiators: Scheraga with the conventional ECEPP and the EPEN concept; Lifson who conceived the CFF and started the optimisation with Warshel and later Hagler; Boyd who also had a quadratically convergent minimisation working in 1968; Kitaigorodskij who pioneered work on molecular crystals; Williams who improved it; Hopfinger[112] who quite early aimed at drug design; and all the rest of us. They quite simply want a black box and do not care what is inside.

It is my intention to help also such colleagues, by introducing to them the many possibilities that exist, and by pointing out the merits and drawbacks of each approach.

Another category is formed by those, of many denominations, who do not shy at the prospect of having to modify a program, learn about different minimisation methods, or choose and improve on their own potential energy functions, to further their research. Such colleagues have often approached me, with constructive criticism, with outright help, with pleas for or invitations to collaboration. The most common question has always been: "Where do we get a potential energy function?" This is the category I mostly have in mind when writing this chapter and, indeed, the entire book.

A third category is a tiny one with which there is little or no communication. It is made up of som of those who are in the business themselves. They give few or no references to work outside their own. In the whole book, they will be interested only in this chapter, and of that only in special sections.

The fourth category is the still extant large population of chemists who believe that everything with a theoretical tinge is quantum mechanics. That category will not even notice the appearance of this book.

3.2   Plan

As stated above, this review cannot be complete in coverage. I do not want it to be overly secterial or fragmentary, so I have chosen to organise it by scientific groups or individuals. This approach breaks a strict chronology, but gives the advantage of presenting the developments from some of the leading groups in a logical order, so that prospective users may see what the different methods have to offer and what they cannot or cannot yet do.

In cases where reviews by other authors give this information, I shall refer to them rather than attempt to rewrite well-written expositions.

It is hopefully evident that I was introduced to the consistent force field through repeated visits to the Lifson group. My selection may therefore be biased, but, so I hope, only to the extent that I put great emphasis on the original and basic ideas of Lifson, as they were expounded in his first paper with Warshel[151]. It has surprised me that no other group than the original and my own has taken to that approach.

It had been in my mind to organise the review according to classes of substances. I found that it would be tedious also for myself to read page after page of alkanes, then a heteroatom or two, etc., so I chose the plan outlined above.

3.3   Historical background

The history of a priori machine computation of molecular conformation has been recorded by the early reviewers[276,69,279], and need not be repeated here; I just wish to make a few points.

Right from the beginning[107,276] the importance and difficulties of parametrizing non-bonded interactions were realised, and the special problems of 1-3 interactions were discussed by Hendrickson in

1961[107]. Already Westheimer[276] noted the importance of electrostatic contributions.

The minimisation and pseudo-minimisation methods used by the pioneers were summarised before[190,p.128].

Although it is mostly overlooked, application of potential energy function methods to conformational analysis occurred in inorganic chemistry, more specifically in coordination chemistry, about as early as in organic chemistry[161,52]; see also chapter 4.

3.4   CFF and variants

In this section I shall trace the development of the consistent force field as made by Lifson with various coworkers, and by some of us alone or with others.

3.4.1   Lifson and Warshel

The consistent force field was introduced in 1968[151] with the optimisation of progressively complicated potential energy functions on 94 observables of 10 n-alkanes and cycloalkanes; the observables were conformations, frequencies and excess enthalpies. The best potential energy function, which became a model for subsequent work in the group, was a modified Urey-Bradley type with quadratic bond, angle and geminal or 1-3 terms, linear angle (only CCC) and geminal terms, a Pitzer term for torsion, and Lennard-Jones 12-6 plus Coulomb terms for 1-4 and higher non-bonded interactions. One of the remarkable features of this potential energy function was that properties of the medium-ring cycloalkanes, also cyclopentane, were reproduced. A drawback was that in a MUBFF there is by necessity strong correlation between angle and geminal terms. 26 parameters were employed.

Introduction of angle-torsion crossterms improved the vibrational spectra at the cost of two additional parameters[264].

The next big step came in 1970 with optimisation on crystal properties[266]. The new potential energy function had the same form as the previous, and many parameters for intra-molecular interactions had changed very little. Angle-torsion cross terms were added, and a Lennard-Jones 9-6 form was used. For some interactions, different parameters were used for methyl and methylene groups. Altogether 29 parameters were employed.

3.4.2   Warshel

Rounding off his work on alkanes to still greater perfection, Warshel[267] introduced anharmonicity into the expansion of the potential energy, taking five in stead of the usual three terms. The potential energy function had a completely new and, so I judge, rather ingenious, form: Morse functions for bond stretching, intrinsically coupled with angle bending; 9-1 power terms for geminal interactions; harmonic angle terms; Pitzer torsions; angle-torsion cross terms; and Lennard-Jones 9-6 plus Coulomb non-bonding terms. Altogether 31 parameters were used for that work.

This endeavour probably represents the ultimate one should try to achieve towards mathematical complexity in potential energy function methods, and the programming effort must have been immense. As far as I am aware, nobody has extended the work to other molecules which should be easy in principle once the coding has been done, but very costly in the number of parameters.

3.4.3   Amides

Calculation of secondary structure and thermodynamic properties of proteins had been Lifson's goal when he began to apply matrix algebra to molecular structure.

Soon after the first consistent force field papers a potential energy function for amides and lactams was optimised[265], a MUBFF with Lennard-Jones 12-6 and Coulomb terms, and out-of-plane terms on the trigonal carbon and nitrogen atoms of the amide group. 54 parameters were used, which number was increased to 82 in a subsequent work by Schellman and Lifson[234], even though most θ_0 were locked to 120° or the tetrahedral angle, and most F' of the MUBFF through $F' = -F/10$.

3.4.4   Hagler and crystals

During the following decade many papers on amide structure appeared from the group around Lifson. They show a progressive deviation from the consistent force field concept and methods:

(1) The internal molecular structure either is treated with parameters unchanged from the earlier works[93] or is kept stiff[94], while non-bonded parameters are optimised on structural and thermodynamic data of crystals.

(2) Lattice summation is performed only in real space at large cost in computer time, and cut-offs are of necessity chosen rather arbitrarily.

(3) ab initio calculations with GAUSSIAN-70 and even rather large basis sets are used[95] to obtain insight into charge distributions and electrostatic interactions.

In the first work in this series[93], a very simple modelling of
the hydrogen bond was used: no other terms than van der Waals and
Coulomb interactions are needed. The idea was not new; Coulson and
Danielsson[54] emphasized the essential electrostatic character of the
hydrogen bond in 1954, and Poland and Scheraga[208] used fractional
atomic charges in calculations on peptides in 1967; but in the first
paper [93] of the series it is used with a consequence not found elsewhere. The same methods were used in a triplet of papers on
carboxylic acids[153,97,98].

Lifson has recently published a review of the basis of the consistent force field and the work of the group up to about this
time[155].

3.4.5   QCFF/PI

Influenced by Martin Karplus, Warshel developed[268,269,270] a version of the consistent force field for use with π-bonded molecules,
in which the σ skeleton is treated with the usual CFF techniques,
whereas a semi-empirical model is used for the π-bonds: an SCF
Pariser-Parr-Pople-type program is integrated in the CFF system, and
its parameters are optimised as the other CFF parameters.

Once again Warshel introduced a qualitatively new development. In
principle, the consistent force field deals with one electronic
state; in the QCFF/PI π-excitations enter, and excited states can be
studied with methods developed on data for the groundstate plus a
few parameters of spectral origin.

Although one may have reservations about the strict separation of
σ- and π-bonds in molecules of no symmetry, there is no doubt that
Warshel's extension works. He has reviewed it himself[271], giving a
very clear introduction.

3.4.6   Ermer

Molecules containing π-bonds were also treated in a more conventional way in the consistent force field framework. Ermer and Lifson introduced[71] the formalism of the generalised valence force field, GVFF, in place of the MUBFF, but still augmented with expressions for non-bonded interactions. A potential energy function with 39 parameters was optimised on 131 data from structures, vibrations and thermodynamics of 23 molecules.

The potential energy function contains terms, all of them harmonic, for bond stretching, angle bending, out-of-plane bending; stretch-stretch, stretch-bend and bend-bend interactions; Pitzer torsional terms; and Lennard-Jones 9-6 terms for non-bonded interactions. It has been used, seemingly without changes, in a number of studies on alkenes and cycloalkenes; the work is discussed at great length in a monography[73]. An earlier and shorter review[72] is still a very good exposition.

3.4.7   Super-CFF

Lifson never left the basic concept behind the consistent force field, as may be seen in two papers from recent years, grown out of an immense amount of computation.

One[96] deals with only one molecule, tris-(tert-butyl)-methane. This overcrowded molecule has very long C-C bonds and other anomalies, and its structure cannot be reproduced properly with any of the previous consistent force fields of the Lifson group. Great effort was put into optimising both a MUBFF and a GVFF on the structure, but in spite of the use of Morse potentials and altogether up to about 30 parameters with limited success. Other potential energy functions, of different constructions and with fewer parameters, give better results, see sections 3.9 and 9.2.

In the other paper[154], a completely renewed potential energy function for alkanes and cycloalkanes was optimised. In form it is a GVFF with Morse functions for C-C and C-H bond deformations, and with numerous cross terms. For non-bonded interactions, Lennard-Jones 9-6 plus Coulomb terms were transferred from previous work[266].

Parameters of all other terms were optimised on 24 conformational and 64 vibrational data for methane, ethane, propane, isobutane and n-butane. Altogether 24 parameters were optimised, out of the total number of 31.

This paper represents the most penetrating analysis yet performed of structures and spectra of small alkanes, and it is imperative for serious students of potential energy functions employed in conformational analysis. Here is a set of functions, with parameters, which reproduce very well observations on methane, cyclopentane, normal alkanes, and, mirabile dictu, tris-(tert-butyl)-methane.

For some reason, the Lifson group, as many others, prefer the r_g to other structures as basis for their optimisations rather than a structure type which is independent of temperature as the potential energy function is; see also section 2.2 and chapter 8.

The reproducibility and reliability of calculations is so high that the differences from one structure type to another are significant, most of all for the C-H bond, on which Lifson and Stern put great emphasis.

3.5 Karplus

Karplus has been interested in conformational analysis by potential energy functions for many years; see examples in section 3.4.5.

Another example is a paper with Gelin[80] on acetylcholine in which they also describe the use of a version of the consistent force field program. They develop a parameter set for a MUBFF by fitting, not optimisation. An important detail is emphasized: minimisation under relaxation in all degrees of freedom is imperative for mean-

ingful results when very flexible molecules are treated. This point
is often overlooked by authors with less insight.

The Karplus group recently published[35] the results of a major
effort: the program CHARMM. It is well described in the long paper;
I shall just mention that it is designed specifically for calculations on macromolecules of biological interest, that it uses several
methods of energy minimisation (including a modified Newton which
like ours avoids trapping in saddle points), can perform molecular
dynamics, and many types of analysis of the results. The potential
energy functions are also given in the paper, with extensive tables
of parameters.

CHARMM is the first program system published which deserves the
designation "molecular mechanics", in the sense that it treats
static, kinematic and dynamic properties. All other programs available are much more limited in scope. A few warnings for the
uninitiated are in place: The more complicated a system is, and here
I mean only modern well-structured and well-documented systems, the
more attention is required to maintain and operate it, and that cannot be left solely to the computer people. Also, know-how does not
come by itself or on a tape. For many prospective applications,
potential energy function parameters, or even the functions themselves, are lacking, just as for the simpler systems. Parameters
must be found by trial and error, as is usual, or, preferably, by
optimisation, which cannot be done in CHARMM; a program of the consistent force field family is necessary.

3.6 Kollman

Weiner and Kollman describe a program, AMBER[273], which is intended for work on proteins. Therefore it uses comparatively simple
potential energy functions, and stored data for building blocks of
large molecules. It belongs in a far-off way to the CFF family, as
it has grown out from the Lifson-Warshel-Levitt programs via the

Gelin-Karplus. The program is modern in intention, but many details are rather primitive: input seems to be cumbersome when compared with the other consistent force field dialects using variants of a method originally invented by Levitt; energy functions are not easily changed; the minimisation criterion is unreasonably liberal. The program does not provide for optimisation of potential energy function parameters.

3.7 Schleyer

Among the early workers in this field we find Schleyer. From a large number of publications I have selected only one[70] for mention here: It is a critical review from 1973, which brings a then new parameter set for calculations on hydrocarbons, a comparison of this with other sets, and important discussions which everybody interested in the field should read.

A detail of high importance to organic chemists, the calculation of heats of formation, is included, in a simple pragmatic way, as a correction to strain energy based on group contributions rather than through statistical-mechanical methods. This approach is followed also by other groups and is not only the easiest, but also the most accurate in the general case. Theoretically better founded methods may be devised, but they are only successful within series of closely related compounds.

The parameter set is interesting in that the geminal interactions are treated through elaboration on angle terms. A cubic function is used with, in principle, different parameters for an angle deformation according to the kinds of non-angle-defining ligands to the carbon atom. In practice, the differences are small or zero. Apart from this, the usual harmonic, Pitzer and Buckingham terms are used.

3.8   Mislow

Also Mislow and coworkers are organic chemists who have done conformational calculations, but of a quite different nature. In
addition to static structure, they are interested in such details of
dynamic stereochemistry as can be studied with NMR; I have selected
two examples.

The first[8] is analogous to TTBM (see sections 9.1 and 9.2): tris-
mesitylmethane. A potential energy function of Allingers is expanded
by trial and error fitting to structures of two cyclophanes and tri-
phenylmethane, and by non-convergent energy minimisation with a
constrained pattern search technique the conformational surface for
flipping of the aromatic rings in trismesityl-methane is explored.
This is one example of the mechanism of an enantiomerisation being
charted by conformational calculation.

The second example is a series of papers[114,115,116], in which the
complicated but fascinating stereochemistry of tetraarylmethanes and
-silanes is explored.

3.9   Bartell

One of the best potential energy functions available for alkanes,
the MUB-1, was published by Bartell and coworkers[121] in 1967 and
later updated[76] and widened in scope[60]. Large deviations from standard unstrained geometry are accounted for with relatively few
parameters[15,217]. All potential energy functions of the group are of
the MUBFF type supplemented with Lennard-Jones and Buckingham terms
for non-bonded interactions. As an interesting detail, 1,3 or geminal interactions are included among the non-bonded interactions,
which probably to a large extent accounts for the precision. The
MUB-2 from 1976[76] with 25 parameters is certainly among the best
available; as other potential energy functions of high precision it
contains a number of cross terms.

3.10 Boyd and variants

Boyd's program[29] from 1968 is very well described. It has been used by many other groups which may also indicate that it is easy to use and to modify, and that the original parameter set was reasonable.

The program appeared at the same time as the original consistent force field, and they have much in common, as well as basic differences. One is that Boyd's program lacks optimisation, another that second derivatives are calculated numerically which is more time consuming than doing it analytically.

This program seems to be the only one, besides our version of the consistent force field, to be applied to coordination complexes.

The important developments came from two Australian groups[40,248] and an American[59]; it was necessary to expand the program to include more atoms and interactions, and to fit (by hand) larger parameter sets. A full documentation of one version is available[58].

Boyd has of course also developed his program further[30].

3.11 Altona and Faber

In 1977 a rather detailed description[74] of the program UTAH5 was published. It has grown out of UTAH[6] which again was a modification of Boyds' program[29]. It can perform much the same tasks as most other major programs; optimisation, as usual, is not available. Two features are worth mentioning: the built-in range of potential energy functions is probably the widest to date (though it seems that the Morse function is not included); and minimisation is performed by a Newton algorithm modified differently from the usual, in that Eckart conditions are imposed on the cartesian displacements. This method should give faster convergence than the generalised inverse method, and should not risk trapping the minimisation in an inflection point on the conformational surface. The program is well described in the paper, but it can not be seen whether its input requirements are friendly to the user.

3.12   The Delft group

A group in Delft has been very active in the field for about a decade. They seem to have stuck to hydrocarbons, but have treated all classes: alkanes, alkenes, aromatics. The group has published a short desription of their program[88]. They have dealt with many aspects of conformational analysis: proper minimisation[13], which even today is often neglected; pathways in conformational space[14,263]; comparison of potential energy functions[89]; chemical equilibria[203]. A range of different parameter sets from different sources are used in the papers from this group.

3.13   White

Two reviews published in a series of very high standard require some warnings against too uncritical reading. Beagley[21] introduces his own shorthand for potential energy functions of most groups. He seems to have misunderstood the basic principles of the consistent force field. White[277] seems to have selected his key references too carelessly; his discussion of non-bonded interactions lacks precision; he does not mention the convergence problem in crystal calculations; he is confused about the distinction between force constant and potential energy function parameter; his treatment of transition states is nonsensical; and remarks about Newton minimisation are simply wrong.

White reviews his own parameter set, which later has been corrected and extended[28] to ethers.

3.14 Kitaigorodsky

Kitaigorodsky was among the first to emphasize the importance of non-bonded interactions for molecular conformation[134,135]. A monography[136] was published in 1973; chapter 7 of the book is a reasonably complete and balanced review of the application of potential energy functions to conformational analysis. Another book[137] has recently appeared.

Kitaigorodsky gives non-bonded interaction parameters[136,p.170] for use with Buckingham functions on aromatic systems. He also gives a rather special potential energy function of the Buckingham type for peptide work, a "universal potential" with just one parameter[135,136, p. 388] for each interaction type. This set has been used also by Rao et. al.[212] in early calculations on saccharides.

3.15 Allinger

One of the two most productive groups in this field is Allinger's. Also, Allinger's programs are used by more people than any other program for conformational analysis. Luckily, Burkert and Allinger in 1982 made a reviewer's task extremely easy[45]. Their book gives a splendid review of their own work, and they do not forget to mention other workers. They cannot have, of course, the same familiarity with other people's methods as with their own, and some problems may therefore escape their attention.

The book provides a good introduction for the uninitiated but it should not be used as the only text. For anyone who wants to begin conformational analysis of a class of small molecules it is advisable to read chosen sections, but the reader should be aware that the book, as is quite natural, is far from complete on subjects outside the authors´ own preferred classes. For example, very little information is given for amino acids, amides and peptides, for which classes other reviews are available, although not so recently.

3.16 Osawa

At least one group has got many programs running, including Allinger´s and members of the consistent force field family[197]. This gives an opportunity to compare different approaches[198] and also to improve upon them, as it was done for MM2[122]. The group works exclusively in organic chemistry, mostly in connection with synthetic work.

3.17 Scheraga

There seems to be general concensus[45] among people who write reviews on conformational analysis to pass lightly as an elf over the huge field of peptides. Even if this practice were followed, it is impossible within the concept I have chosen to do justice to the most productive group of these decades. Scheraga started early into the field, see reviews[235,236] of the early work of his group, before technical refinements; he never left it; he has introduced new models for potential energy functions, and he is very active also in the experimental field, which aspect is completely left out here.

3.17.1 ECEPP

Scheraga always aimed at descriptions of the secondary and tertiary structure of polypeptides, and was therefore always eager to exploit the ever-growing capabilities of computers.

The degrees of freedom in peptides are so numerous that one has to compromise between completeness and practicability. Scheraga uses mainly torsional angles as the only variables, and torsional and non-bonded interactions as the only terms in potential energy functions. The facts that valence angles easily open and the peptide unit is not rigid mean that much work must go into selecting reasonable standard values for bond lengths and angles and into refining non-bonding parameters which will compensate for the inadequacy of the potential energy function. As an example, see the detailed and lengthy discussion of the development of a parameter set which has been used for many amino acids and derivatives[176]; note also the enormous computational requirements when minimisation is done on crystals without the use of convergence acceleration. Even this big work was completely revised shortly after[177]; it has since, as ECEPP, been used in numerous works; a status is available[237]. In a variant of ECEPP, hydrogen atoms are merged into their chain atoms[62]. Out of these endeavours grew attempts at protein folding[181], which is outside the scope of this book. Also the very important aspect of the influence of hydration upon conformation can be treated under ECEPP[111] and other schemes[182]. Recently, one more updating of ECEPP, ECEPP/2, was introduced[183].

My one reservation about the approach, the stiff units, seems to be shared by Karplus[91].

3.17.2 EPEN

A new model, EPEN, was introduced in 1975[244]. It is attractively simple, and basically different from all other empirical methods. The usual approach is to model a molecule as an overlay of the classical picture of mass points connected by springs and the Coulson picture of atomic cores in a structured cloud of electron density, each picture leading to terms in the potential energy function. Scheraga and his coworkers view matter as a collection of nuclei and electrons governed by three types of interaction: Coulombic between all the charged particles, exponential terms to model electron-electron overlap repulsion, and London attraction between heavy atom fragments; the last term probably being a concession to the oversimplification of the model. Electrons are located in lone-pairs or in bonds, and small molecules and molecular fragments have fixed geometry. EPEN was quickly applied to many systems[42,43], and the hope to get rid of torsional potentials and special functions to mimick hydrogen bonding seemed fulfilled. The calculation of energy in crystals is not documented. Williams has made some tests with EPEN under his convergence acceleration program[283] and mentions that EPEN requires ten times as much time as conventional potential energy functions.

In the long run EPEN was unsatisfactory, especially for properties of liquid water, and the model was changed[246]. In fact, it was made simpler and more logical: the potential energy function is composed of Coulomb interactions between all charged particles, exp-6 terms between all electrons, and no more. Fixed geometry of molecular fragments is still used. Crystal calculations are now safer, as Williams´ program[280] is used. EPEN/2 is parametrized for many classes of compounds, and the number of parameters is kept small despite the necessity of using different type of electrons.

3.18 Ab initio modelling of non-bonded interactions

Many authors have tried to utilise results from large-scale SCF calculations with programs such as IBMOL, GAUSSIAN and MOLECULE to derive model potential energy functions. Such calculations on any but the smallest molecules are so costly that it is out of question to perform geometry optimisation in the general case, yet it is silly to let a big computer run for hours to obtain essentially one number, an energy. This is the motivation for many authors to perform ab initio calculations, often with fairly large basis sets, on small or medium-sized molecules and then use the potential energy functions, fitted from the energy terms, in calculations on large molecules or molecular aggregates.

Two approaches towards a rational use of ab initio data for modelling potential energy functions, which may be employed for conformational analysis just as empirical functions, have been taken. I shall illustrate them by selected examples, and I shall mention only the modelling of non-bonded interactions.

3.18.1 Function fitting to calculated energies

A group around van der Avoird has studied the ethylene dimer since 1975[286] and has fitted potential energy functions for non-bonded interactions to ab initio results; they have been used also to calculate lattice vibrations of crystalline ethylene[272]. Also another group has worked on the ethylene dimer[256].

Clementi and his associates have computed the energy of a water molecule and an amino acid, for all 21 naturally occurring amino acids, and for many positions and orientations of the water molecule[50]. A Lennard-Jones 12-6 function plus a Coulomb term were fitted to the total energy, and the result checked on one of the amino acids[27] and then used to study the interaction of water with lysozyme[211]. Similar work was done for the four DNA bases[239] and for the sugar-phosphate-sugar fragment of nucleic acids[163]; a review of these developments is available[260].

Other authors took up the method and made some detailed studies, mainly on hydrocarbons, using more refined potential energy functions[82,77].

3.18.2 Clementi´s Bond Energy Analysis

Clementi published in 1976 a book[49] which in the main is an early review of his and his coworkers studies on liquid water, but which also presents the first account of a new method of deriving qualitative insight from ab initio calculations.

The idea is to decompose the one-, two-, three- and four-center integrals into one- and two-center terms, put them into matrices, and read directly the numerically more important (bonding) and less important (non-bonding) terms of interaction in molecules and molecular complexes. The method is convincingly illustrated with water and methane plus water. Unfortunately the formulae are not free from errors.

The method was expounded later[53], this time more at length, but not altogether transparent. The paper contains a wealth of ideas and Ansätze; the hope of using one ab initio calculation on one conformer of a sufficiently large molecule to derive non-bonded functions and their parameters certainly triggered us to perform probably the largest ab initio calculation to date on a sugar[170]. Qualitatively, the work was a success; we even got a theoretical basis for the then new form of some of the non-bonded interaction terms of our new potential energy function, PEF400[171]. Quantitatively, we were disappointed, as the energies were by and large one power of ten too high (numerically).

The last word has not yet been said: alternative decompositions of both overlap and energy integrals have been tried by a student[145], but that work has not been finished.

3.18.3 EPEN going quantum

The large sets of ab initio energies for water calculated by Clementi and coworkers, and equally large sets of data for methane plus water and formaldehyde plus water, were used by Beveridge and coworkers to fit parameters of the EPEN/2 potential energy function (see section 3.17.2). The conclusion was that whether this development will succeed in the long run depends probably on how costly the method becomes when large solute molecules are tackled; the start has certainly been promising.

4   Applications: Coordination complexes

> There is one thing worse than fighting with
> allies, and that is fighting without them.

> Churchill, quoted by Alanbrooke

It was mentioned in section 3.3 that conformational analysis was performed on coordination complexes as early as on organic compounds. It is true that Mathieu[161] was too early to set the train moving in this direction, and the development was of course hindered by the ongoing world war; but Corey and Bailar[52] got it on the rails, and opened new vistas.

Studies of coordination complexes were a major activity of our institute for a century, and it was evident that we should embark also on calculational methods of conformational analysis when the possibility arose [190,Preface].

This chapter will be a review of work in our own group, but will not fail to mention important work of others. A comparative review, larger in scope, will be timely after the next two development steps which I think will be proper optimisation and a priori calculation of stability constants.

The most important aspects of our work up to 1980 has been reviewed before[216], unfortunately not entirely free of misprints: on p. 220 a "T" is missing in the formula for E_{vib}, a "1" before a minus sign and a "/" before a k in that for S_{vib}, and "lbn" on p.221 should be "ibn". The short review also brings original material, notably pictorial representations of conformers of $M(tn)_3$ and $M(ptn)_3$, a revised table of thermodynamic data, and consistent force field calculations of vibrational spectra of $Co(en)_3$ and $Co(tn)_3$. (For abbreviations see Table 4-1.) The approximations behind our statistical-mechanical calculations, which are the same as used by most other groups, are also listed there.

Table 4-1: Abbreviations of amines

---

aa	diamine
bn	2,3-butanediamine
bnta	2,3-butanediamine-N,N,N´,N´-tetraacetate
daes	di(2-aminoethyl)sulfide
en	1,2-ethanediamine
enta	1,2-ethanediamine-N,N,N´,N´-tetraacetate
ibn	2-methyl-1,2-propanediamine
ptn	2,4-pentanediamine
tn	1,3-propanediamine

4.1   An early consistent force field

The work on adaptation of an early version[265] of the consistent force field program to coordination complexes began in 1969, when I first joined the Lifson group. I had the program working on our complexes as far as chemical formula interpretation and minimisation, but optimisation on the rather large molecules would require very tedious and also costly handling of the machines then at our disposal in Israel and Denmark. Therefore optimisation was limited to diamines in the gaseous state and in isolated chelate rings, in order to gain some experience. The results were so encouraging that we tried to publish them, but the paper was refused by Inorganic Chemistry. In consequence, these first optimisations on coordination complexes are reported only in my Ph.D. Thesis[213]. The potential energy function was a MUBFF similar to what was then used in the

Lifson group; optimisation was made on Saitos "average Coen ring"[230], the then new x-ray data from Ibers and coworkers[225,226] on $Cr(en)_3$ complexes, vibrational spectra of normal and N-deuterated en[229], and my own spectra of the complexes[213].

Despite all the shortcomings, the results were as good as in contemporary work on organic compounds. Figure 4-1 shows the structural details; the vibrational data for all four "compounds": en, en-d$_4$, Coen, Cren, showed deviations from + 183 to - 91 cm^{-1}, with an average of +3 cm^{-1}; in most regions the spectra were better reproduced than with another potential energy function developed later without optimisation[216].

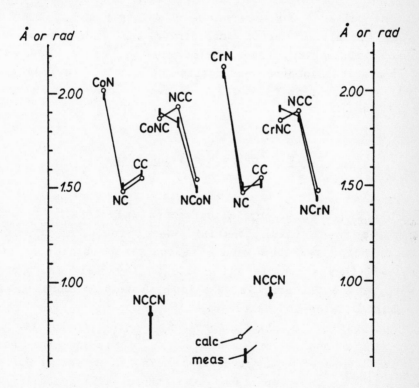

Figure 4-1: Early optimisation on single chelate rings

4.2   Australian and American groups

While the consistent force field was being developed other groups had been active. Four series of works shall be mentioned here.

4.2.1   Gollogly and Hawkins

One of the first attempts at a development of the ideas proposed by Corey and Bailar[52] was undertaken by Gollogly and Hawkins. They used a very primitive method considering only tiny parts of the energy, no minimisation, wrong thermodynamics, and few degrees of freedom. A number of papers were published[84,85,86,87] in the journal which refused consistent force field work.

4.2.2   Sargeson and coworkers

A group of Australian crystallographers and classical coordination chemists found that potential energy function calculations might be of help, and they were wise enough to obtain one of the best programs then available and modify it for their own purposes, which were, and still are, calculations on Co(III) complexes with amines, aminoacidates and halides as ligands.

Their adaptation of Boyd's program[29], see section 3.10, is described in a number of papers[248,39,40,68], and the results obtained are compared with structural data in the same and other papers[247,81,249]. In comparison with these achievements, which represent both programming efforts and painstaking selection of potential energy function parameters, it is a minor blemish that they left out long-range non-bonded interactions, for reasons of computer economy. It is a pity they did, as they were led to wrong conclusions[81] as to relative stabilities of conformers of $Co(tn)_3$[189].

The early work of the group is dealt with in a review[40] which should be read by anyone who wants a start into this field.

4.2.3   Brubaker and coworkers

Brubaker and Euler[37], and later Brubaker and Massura[38], used the same program in studies of complexes of a picolyl substituted diamine. Note their honest apology for using interaction parameters from different and unrelated sources; it is sound and courageous.

4.2.4   DeHayes and Busch

Another adaptation of Boyd´s program was done by DeHayes[58], as mentioned also in section 3.10. As far as can be seen, this installation was done with fewer modifications, notably without built-in errors in the treatment of non-bonded interactions. The applications were confined to amine complexes and centered on the study of intra-ring strain[59].

4.3   The Lyngby group

Two series of works were done here, using two rather different programs.

4.3.1   Niketić and Woldbye

The first works in this group were based on Wiberg´s program[278] which was adapted by Niketić and applied to Co(tn)$_3$ complexes with six-membered rings[186,188] and to complexes of five- and six-membered rings with methyl substituents[187]. These papers served to develop a nomenclature for conformers, to select and modify potential energy function parameters, and to gain experience in the calculations. The results cannot be considered final, as the method of minimisation used was the steepest-descent which is not convergent and in principle never reaches a minimum.

4.3.2   Rasmussen and coworkers

During the seventies, our computer centre was always under change: from IBM 7090 to 7094, to 7094 II, to a small 360, to a bigger, etc. In consequence, the development of a rather big program was a very time-consuming task. The consistent force field program I brought home in 1970 was completely rewritten by Niketić and myself, with the exception of one single subroutine (Davidon-Fletcher-Powell minimisation), and much was changed, such as the Newton minimiser. This development has been described before[190].

The new program, which went as far as minimisation, was applied to all conformers of Co(tn)$_3$ and some of Co(ptn)$_3$[189], to all of Co(en)$_3$ and Co(bn)$_3$[191], and to all of Co(ibn)$_3$ and some of Co(bn)(en)$_2$[99]. In all of these works, and in the following to be mentioned, the parameter set of Niketić and Woldbye was used, with a slight change. This led to disaccordance with x-ray structures for some bond lengths and angles, up to 0.06 A and 4° in the worst case, but dihedral and torsional angles were generally well reproduced. Much insight was got into conformer energy differences: ob-lel, chair-lel-ob; and into the influence of methyl substituents: mer-fac, and bn vs. ibn.

In parallel to this work, the program was developed to include vibrational analysis[190], chapter 6 and statistical thermodynamics[100], and an implementation of a previous optimiser[190], chapter 7 which was later found unsatisfactory. Interfaces to the vector plotter programs MONSTER, ORTEP and PLUTO were written, and a very useful application to normal coordinate analysis was made[214].

These developments led to proper calculation of stability differences[100,216,193]. The last of these papers rounds off the long series of studies with a new treatment of $Co(en)_2(tn)$ and $Co(en)(tn)_2$ and supplementary thermodynamic calculations on the previously studied systems. In addition, the first attempt to calculate stability constants in this framework was made.

Despite the non-optimised parameter set, the calculations gave reasonable postdictions and predictions of equilibrium distributions of conformers and to some extent isomers; good agreement is found for $Co(aa)_3$ with aa = en, rac-bn, meso-bn[193].

A technique for following conformational interchange, originally developed for saccharides[168], was perfected in a study[192] of the ring-flip mechanism in $Co(en)_3$. The path in conformational space was calculated and drawn, the barrier found, the rate estimated, and the two-dimensional normal coordinate drawn in stereo.

Our work has been reviewed at intermediate stages by ourselves[216] and by Saito[231,232].

Since 1980 we have not worked on coordination complexes, but have rather striven to perfect our optimisation (see chapter 7), and expand to calculations on crystals (see chapter 11). When we consider ourselves experts in optimisation (and know-how is accumulating, see chapters 9, 10 and 6), we shall return to coordination complexes, which are very costly to optimise on.

4.4   Other groups

While we worked in Lyngby, others had not been idle. Not unex-
pected, the other groups used either Boyd´s program or ours.

4.4.1   Pavelčik and Majer

A Slovak group uses Boyd´s program in DeHayes´ version[201] to analyse strain in the rings of the enta (or EDTA) complex of Co(III).
Their parameter set is made up from many different sources. A later
work deals with the Co(III) complex with bnta (or DBTA)[202].

4.4.2   McDougall and coworkers

A South African group also began calculations with Boyd´s program. They got it in Snow´s version[248], and applied it to stability
differences between complexes of Ni(II) with five- and six-membered
chelate rings[166,26] and other thermodynamic calculations[102,103]. We
have earlier commented[193] on the three first papers, and there seems
no reason to revise our remarks that the method leaves out the
majority of the non-bonded interactions (for which there is no
excuse in 1978 and later), that the minimisation criterion is very
liberal, that calculated energies are compared with measured enthalpies with no account of the internal motion. Altogether, their
method has no interest as a modelling tool.

4.4.3   Australians

The two Australian groups have merged their crystallographic, computational and NMR skills in a study[101] of conformations of a Co(III) complex of ptn in crystal and solution. Such studies have not been common, but should be encouraged. It is pertinent to remark that calculations are performed on molecules all alone in the world, as it were, and measurements either on crystals or on solutions usually with high ionic strength.

4.4.4   Laier and Larsen

For a change, this group used our program in a study[143] of amine and thioether complexes of Co(III) explain why only one isomer of Co(daes)$_2$ exists. Apart from this, the paper is interesting for a different parametrization which includes sulphur with and without lone pairs. The easy adaptability of the program is neatly demonstrated.

Laier and Larsen used three-parameter Buckingham functions composed from Allinger's two-parameter Lennard-Jones functions[3]. A quite unnecessary complication, as the program would work equally well with Lennard-Jones functions.

4.4.5   Bugnon and Schlaepfer

A Swiss group took over the consistent force field program I left after a short stay as guest professor in Fribourg. Much to my surprise they managed to use it in a study of Jahn-Teller distortion in Cu(II) complexes[41].

4.4.6   Tapscott and coworkers

One more group has used Boyd´s program in DeHayes´ version, in a study[109] of isomers of Co(bn)$_3$, where they confirmed our earlier findings[191,100]. The group chose to concentrate on experimental studies, and they have produced some remarkable results, which have been[78] or will be used in the checking of calculations[191,193].

Schemes lightly made come to nothing,
but with long planning they succeed

Proverbs XV, 22

The wish to study conformations of saccharides came in 1973 when I heard a lecture by Otto Westphal on bacterial endotoxins. Since then one of my main goals has been to find the conformation of "the endotoxic principle", Lipid A. This is a disaccharide[227] consisting of two glucosamines, heavily substituted with long fatty acids, ester-bound and amide-bound. Therefore[215] we began developing a potential energy function for saccharides, starting with glucose.

The development of potential energy functions and parameter sets has been described in detail in the original papers, so I give here just references, and a summary of our results, supplemented with comparisons with the results of later experimental work from other groups. The large majority of calculations on saccharides reported in the literature are limited in method, as subunits are kept stiff, and are not mentioned here.

5.1   The glucoses

In our first glucose paper[132] we described the development of FF3, later called PEF3, through a comparison of different types of potential energy functions and trial and error fitting of parameters. In retrospect, here is a warning against merging parameters from different sets, and using them on molecules for which they were not intended.

PEF3 gave quite a good fit of bond lengths and valence angles, and of those torsional angles that are not dominated by hydrogen bonding in the crystal. Also the equilibrium distribution on the two anomers was well represented.

The simpler PEF300 was introduced[167] to test a hypothesis that torsional terms around single bonds in saturated compounds are not needed provided non-bonded interactions are treated adequately. This was indeed so.

Subsequently, PEF400, see section 5.4, was applied also to glucose[172,215] The main results were almost the same as before, except for a perfect match of the anomeric ratio. Details of ring conformers and their relative energies were also discussed[215].

A later work[218] dealt with two questions: (1) what effect does a change from a three-parameter Buckingham to a two-parameter Lennard-Jones function have, nothing else being touched? (2) when Coulomb terms are included, what is the effect of using the same charge for all carbon atoms, also the anomeric?

The answer to the first question was that the simpler function, with fewer parameters, give as good results as, in fact slightly better than, the more complicated one. The answer to the second was, somewhat to my surprise, that there are no effects on conformations, but that different carbon charges cause a slightly wrong estimate of the anomer equilibrium ratio. Altogether welcome conclusions when one wants to cut the number of parameters.

Since our calculations were made, new x-ray data have appeared. Jones et al. solved the structures of peracetylated α- and β-D-glucopyranose[125]. A comparison with calculations using PEF300 is of interest. Bond lengths and valence angles are reproduced with the same precision is in the earlier comparisons, but the ring is more flattened in the acetylated compounds than in the unsubstituted. Calculations are therefore indicated when we have obtained optimised parameters for the ester group.

Sheldrick and Akrigg determined the structure of 3-O-methyl-α-D-glucopyranose[242]; it is somewhat unusual in that the C5-O5-C1 angle is as open as 115° and C4-O4 as short as 1.396 Å. These anomalies are not fitted by our calculations. The C4-O4 anomaly may have to do with an intermolecular hydrogen bond; here is therefore a candidate for calculation with our new crystal program and a potential energy function like PEF401 or PEF304 of section 9.2.

In all three studies the conformation around C5-C6 is ga, the one calculated to have lowest energy[215]. In glucose-6-phosphate[130] the conformation is g´g, that calculated to have second lowest energy[215].

A calculation on α-D-glucose surrounded by 10-12 water molecules has appeared[139]. As expected, torsions become more realistic, especially the exocyclic C-C-O-H torsions. The potential energy function is much more complicated than any of ours, as it has originated from Allinger´s MM2; it even contains individual terms for C1-O1, C1-O5 and C5-O5 bonds. In view of this, the modest accuracy in bond lengths is surprising.

5.2   Maltose

Before going straight on to Lipid A, we wanted to check disaccharides on the way, and chose first β-maltose, 4-O-α-D-glucopyranosyl-β-D-glucopyranose, for which structural data were available. We found[171] all conformers in PEF300: there are four; those of lowest energy compared well with crystal conformations.

We also found the path in conformational space between the two most populated conformers, the barrier height, and the rate of interchange, which is fast on the NMR time scale. The weighted average of a specific proton-proton distance across the glycosidic linkage agreed with NMR measurements.

The conformers were redetermined when PEF400 was ready[172]; a detailed discussion of the changes, and comparison with more crystal structure determinations, is available[215]. The very slight conformational changes caused the minimum valley in conformational space to bracket the crystal data more closely.

A thorough x-ray investigation of β-maltose octaacetate[34] lends support to our calculations. The crystal conformation is close to our middle conformer in the minimum valley, see Table 5-1. The C5-C6 conformation is g´g in the reducing sugar and alternateley ga and ag´ in the nonreducing, confirming our calculated possibilities. For ease of comparison, I keep our previous nomenclature[215] here.

Table 5-1: Maltose conformers

C1´O4C4	φ	ψ	H1´---H4	PEF	min no	ref
117.5	-29	-36	2.35			34
114.0	-21	-24	2.17	300	1	215
116.9	-18	-31	2.38	400	2	215

As to the H1---H4 distance, note that a measurement in solution will correspond to an average over three conformers, or three families of conformers, considering the exocyclic torsional degrees of freedom.

In connection with the crystallographic work a primitive conformational analysis was carried out, with results inferior to our earlier ones[168,215]; let me just recapitulate that other crystal conformers are clustered at another of our calculated conformers.

Tvaroška made a simulation of maltose in various solvents[262]. Based on coordinates gotten from us, he estimated the influence of changing solvent properties on relative conformer populations. One interesting result is that in many solvents a more balanced population is expected than what is calculated for the naked molecule. Our conformers 3 and 4[168], for example, should be the most populated ones in water. An obvious shortcoming of Tvaroška´s model is that hydrogen bonding is not taken into account, such as it is, though not sufficiently well, in our potential energy function PEF400. In this, the former conformer no. 3 should be preferentially populated, but no. 4 should not be present[215]. Let me add that we expect a high barrier between the no. 1-2-3 family and no. 4.

Tvaroška hints[262,p.1894] that other stable conformations may exist. He is right to the extent that, just as for glucose, each conformer described is a representative of almost a continuum of closely related ones, all with the same or almost the same glycosidic torsional angles ϕ and ψ, but with different torsions around the exocyclic C-O bonds. I find it hard to imagine the existence of more gross conformers, with greatly differing ϕ and ψ.

5.3 Cellobiose

For the next saccharide molecule we chose β-cellobiose, 4-0-β-D-glucopyranosyl-β-D-glucopyranose, which also was structurally well-known, and for which we expected slightly more conformational freedom because of the β configuration of the non-reducing ring. We found[169], again with PEF300, six conformers, two of them close to crystal conformations. We also found the path linking the two, the transition rate, and the weighted average H-H distance across the glycosidic linkage, which agreed with NMR results.

The switch to PEF400 caused a very neat bracketing of the crystal conformers by two minima[215]. Also the conformation of the octaacetate[148], which was not included in our first comparisons[169], is close to our global minimum.

5.4 Gentiobiose

Our third disaccharide was β-gentiobiose, 6-O-β-D-glucopyranosyl-β-D-glucopyranose, consisting of two glucose units in the same linkage as the disaccharide in Lipid A. For this work[172], we changed to PEF400. There were several reasons for the change, among them the lack of modelling of hydrogen bonding in PEF300, and a somewhat too small inter-ring C-O-C angle in maltose and cellobiose. The development of PEF400 is thoroughly described elsewhere[171,215].

When the first calculations were made, there were essentially no experimental data available for comparison. Due to the very flexible glycosidic linkage, many minima were found, certainly not all, hopefully all those of low energy. For the same reason, it is difficult to make a good crystal.

The only experimental data to compare with came from the NMR studies of Bock and Vignon, published later[24] with corrections to the data cited by us. They used ^{2}H NMR data to show that octa-acetyl-β-gentiobiose tumbles isotropically in chloroform, and that conformation change around C5-C6 is not substantially faster than the tumbling. This is to be expected, as the dynamic situation must be dominated by the large peripheral groups and the viscosity of the solvent; intrinsic barriers to conformational change around glycosidic bonds are very low, even for 1 → 4 linkages, see sections 5.2 and 5.3.

In this connection attention is drawn to a warning by Jardetzky[124] against interpretation of average values of NMR parameters in cases of more than one dominant conformation.

Bock and Vignon measured ^{1}H and ^{13}C NMR spectra of β-gentiobiose-octaacetate in CDCl$_3$ and performed a conformational analysis from δ, J, T$_1$ and NOE data in conjunction with a primitive calculation[25] on the unsubstituted saccharide. I shall focus on torsional angles in and H---H distances across the glycosidic linkage. Using mapping, not minimisation, and the crude model of stiff monomers and hardsphere interactions between non-bonded atoms, Bock and Vignon concluded that the octa-acetate exists in solution with the same ϕ and ψ as in the crystal, and with two values of ω, corresponding to the g´g conformer found in the crystal, and to the ga conformer.

In their comparison with our calculations, they cited one conformer out of the 25 reported[172] which has a completely different value of ψ, chosen presumably because it has lowest energy and free enthalpy. They overlooked that within 10 kJmol^{-1} fourteen other conformers are found, one of them, at 5.1 kJmol^{-1}, having correct ϕ and ψ, and ω corresponding to the ga conformer; another, at 9.4 kJmol^{-1}, having correct ϕ, ψ $40°$ wrong, and ω corresponding to g´g.

Table 5-2: Conformers of B-gentiobiose in PEF300

No.[a]	ΔG	C1´06C6	ϕ	ψ	ω	H1´---H$_R$	H1´---H$_S$	C5-C6
2	2.25	113.4	47	−116	−177	2.29	3.51	g´g
6	5.54	113.7	−15	−158	−177	2.26	2.62	g´g
8	1.80	113.2	34	−91	44	2.34	3.53	ag´
9	2.95	113.3	32	−90	43	2.34	3.52	ag´
16	0.00	112.6	43	172	−57	2.39	2.68	ga
17	1.60	113.9	20	−93	−52	2.22	3.47	ga
23	2.39	113.3	−11	−170	−57	2.35	2.46	ga
24	2.85	113.4	−15	−162	−58	2.29	2.56	ga
x−ray		113.3	63	−156	−178	2.40[b]	3.12[b]	g´g
PEF300	3.00	113.5	47	−177	−178	2.28	3.50	g´g
PEF400	0.80	115.4	85	−155	152	2.66	3.56	g´g
Bock and Vignon[24]				−170		2.31	3.09	g´g

a The numbering refers to conformers in PEF400[172].
b Bock and Vignon[24] state 2.35 and 3.35.

In order to see what changes the removal of the modelling of hydrogen bonds inherent in PEF400 would bring about, I repeated all calculations[215], using PEF300. The conformer of lowest free enthalpy has hydrogen distances H(C1´)---H$_R$(C6) and H(C1´)---H$_S$(C6) of 2.39 and 2.68 A; Bock and Vignon found 2.31 and 3.08 A in their modelling, but overlooked our results with PEF300.

A closer inspection of the results of the PEF300 minimisation revealed that altogether eight conformers are found within 10 kJmol^{-1}, all having the same order of the two crucial H---H distances; details of their conformations are shown in Table 5-2 together with data from the x-ray structure and from the "gg" (= the g´g) conformer of Bock and Vignon.

Table 5-2 shows that the H1´---H$_R$ and ---H$_S$ distances may not be exceptionally good probes of the whole conformation. It also suggests that other conformers not close to the crystal one may co-exist in solution.

The first crystals of the very flexible substance were prepared and studied by Arène et al.[9]; just afterwards Rohrer et al.[228] completed a full study on succesfully grown crystals. They gave a detailed discussion of all aspects of the structure.

An interesting feature of the crystal structure is that all hydrogen bonding is inter-molecular. In calculations on isolated molecules, hydrogen bonding, if it is taken into account, is necessarily intra-molecular. Therefore the calculated exo-cyclic torsions cannot be expected to agree with the experimentally found, neither when PEF300 (no hydrogen bonding) nor when PEF400 (strong hydrogen bonding) is used. On the other hand, structural features not dominated by hydrogen bonding should be well represented by both.

The coordinates given by Rohrer et al. were therefore transformed to cartesians and used as input conformation with both PEF300 and PEF400. Resultant conformers are shown in Figure 5-1 together with the crystal conformation. The figure, and also prints of relevant bond lengths and angles, show that there is a coordinate error for O6´ in Table I of Rohrer et al.[228] The glycosidic COC angle, torsional angles ϕ, ψ and ω, and the two proton-proton distances are given in Table 5-2.

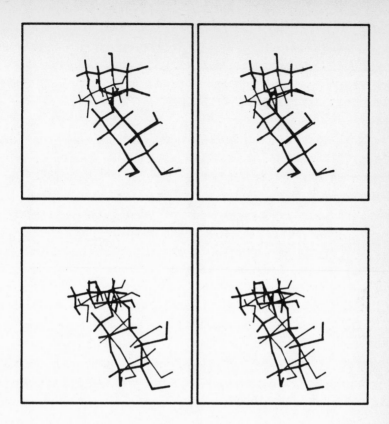

Figure 5-1: β-gentiobiose in PEF300 (above) and PEF400 (below) and
in the crystal (thin lines)

The conformation proved unusually difficult to minimise in both
potential energy functions; there is apparently no minimum in either
PEF300 or PEF400 close to the crystal conformation. The new conformer in PEF300 is very close to No. 2 of the calculations above,
and that in PEF400 is much like the previous[172] No. 4. Figure 5-1
shows that the crystal disposition of exocyclic OH groups is kept in
PEF300, but that a pattern of intramolecular hydrogen bonding has
developed in PEF400, as expected.

Rohrer et al. performed an energy minimisation using only Lennard-Jones functions; they came to a conformer with ψ and ω as in the crystal, but with a completely different ϕ.

In connection with x-ray diffraction on scleroglucan fibres, in which the side chains are β-D-glucopyranosyl residues (1→6)-linked as in gentiobiose, Bluhm et al.[23] performed a series of calculations on gentiobiose. They used a well-established parameter set for non-bonded interaction supplemented with various correction terms, stiff residues, and stiff angles in the glycosidic linkage. No energy minimisation was made, only mapping over a grid in three torsional variables. In this way they derived favoured conformations, 8 g´g, 5 ga and 3 ag´, with respect to the C5-C6 bond.

It is difficult to compare those results with ours; partly because they are not minimum energy conformers, may in fact be far from minimum; partly because only energy, not free enthalpy is given; it is our experience that the ΔE scale may easily differ from the ΔG scale by more than 5 kJmol^{-1}. It is therefore puzzling that the authors state that their results are in agreement with ours. What they have achieved is, qualitatively, the same as we: to have demonstrated that isolated gentiobiose molecules may coexist in quite different conformations.

5.5   Galactobiose

In the same way as for gentiobiose, the conformational space of galactobiose,   6-O-β-D-galactopyranosyl-β-D-galactopyranose,   was explored, using PEF300 and PEF400[224]. Also in this case, on account of the very flexible glycosidic linkage, many conformers were found. There is a tendency of clustering in the case of torsional angles for PEF300 and in the energy for PEF400; each tendency making the situation more homoconformational.

One conformer has been used for in a study[83] of the binding mode of galactan to an immunoglobulin, monoclonal anti-galactan IgA J539.

Docking on a computer, using coordinates of a low-energy conformer in PEF400, and coordinates from an x-ray study at 2.7 Å resolution, gave two possibilities for the orientation of the polysaccharide; fluorescence experiments gave clear evidence for one of them.

5.6   Misunderstandings

Many people have cited our saccharide papers, unfortunately not always with complete understanding. In case we have not managed to explain our ideas and methods clearly enough, I shall here endeavour to correct at least some of the misunderstandings.

Joshi and Rao wrote about flexibility in the pyranose ring[126]. They cited us for not having indicated possible ring distortions or the energies involved, which is just what we did in the paper[132] they cited. They themselves did not perform minimisation.

Dais and Perlin[55] have examined the possiblity of using ^{1}H and ^{13}C NMR chemical shifts to infer relative stabilities of disaccharides. They cited us[172] as stating that gentiobiose is more stable than cellobiose which is more stable than maltose.

There is absolutely no such indication in our published material. Moreover, Dais and Perlin are confused in two respects. One is that they multiply our ΔG values, already in $kJmol^{-1}$, by 4.184. Another is that they judge stability of one compound over another by comparing differences between ΔG values for two out of many conformers of each, and on scales that are arbitrary, in the sense that for each compound the zero is set to the G of the conformer of lowest G.

The basic idea of Dais and Perlin is, however, sound enough, and we have employed it before in a different context: relative stabilities of isomers and conformers of tris-(diamine) coordination compounds, derived from absolute free enthalpies[100,216], see also section 4.3.2. The condition for doing so without further calculations is that the isomers have the same types and numbers of interactions. If this is not the case, rather elaborate corrections

are necessary, as has been shown for complexes of amines of different sizes[193].

Maltose, cellobiose and gentiobiose fulfill the above condition,
and their absolute free enthalpies in PEF400 were found before for
gentiobiose[172], and for maltose and cellobiose[215]. The total free
enthalpies, summed over all conformers and weighted by their Boltz-
mann distributions, are shown in Table 5-3. Results obtained with
PEF300 were given before for cellobiose [169]; for maltose and gentiobiose they have been calculated here. A result for galactobiose is
also given in Table 5-3.

Table 5-3: Absolute free enthalpies in kJmol^{-1}

	mal	cel	gen	gal2
PEF300	882.4	882.3	878.9	879.9
PEF400	796.1	791.9	765.3	

The results are more dependable for maltose and cellobiose than
for gentiobiose and galactobiose. For the two 1 → 4 linked disaccharides, we feel certain that we have identified all important
conformers; for the much more flexible 1 → 6 -linked we cannot be
so certain; the results given in Table 5-2 and discussed above sug-
gest that more conformers of low energy do exist.

We can therefore conclude that in so far as PEF300 and PEF400
give an adequate description of energetics as well as of structure,
the sequence of stability guessed by Dais and Perlin holds, but that
the differences are very small and, for PEF300, are probably not
significant.

The reader should note that all the G values in Table 5-3 are
absolute, on the same scale for three compounds in one potential
energy function, and that the zero of each scale is arbitrary.

Lemieux and Bock wrote in a review[147,p.132] that they were "in sharp contrast" to us[172] (among others), "who have created special functions ... to involve intermolecular hydrogen bonding as a principle force controlling the conformation of maltose". We never did so, and would most probably not have done it in a paper on gentiobiose.

6 Applications: Other compounds

It is ever so with the things that Men begin:
there is a frost in Spring, or a blight in Sum-
mer, and they fail of their promise

Tolkien: The Lord of the Rings

In this chapter I give a short review of the work our group has
done on conformational analysis of compounds not mentioned in other
chapters.

6.1 Amines

The parameter set which we used in used in our many studies on
coordination complexes (see section 4.3) was employed also in an
attempt[220] to reproduce the known conformers of 1,2-ethanediamine
(en) and predict the unknown of 1,3-propanediamine (tn). It was also
our aim to see how far we could stretch the usefulness of the parameter set which was developed for coordinated amines, and does not
include any modelling of hydrogen bonding. Lastly, we wanted to
check with results obtained with PCILO and GAUSSIAN. The conclusion
was that the empirical method gives much better geometry than PCILO
but that PCILO can give crude estimates in the absence of a good
potential energy function parameter set.

The work shows that the parameter set might be quite a good candidate for optimisation. Another approach might be to use a
potential energy function of the form of PEF401 or PEF304, see section 9.2, and optimise simultaneously on gas-phase properties of
amines and crystal data of their ammonium salts with simple anions.
Some exploratory work already done has been encouraging.

6.2   Polynucleotides

Large-scale ab initio calculations on fragments of polynucleo-tides had led to a parameter set for non-bonded interactions[259]; see also section 3.18. We wanted to utilise this set in proper energy minimisation, and composed a potential energy function with para-meters from many different sources[261].

Among the conclusions from this introductory work are that when studying flexible molecules and fragments of this size it is impera-tive to let relaxation occur in all internal degrees of freedom; that a model for nucleic acid backbones should be based on a larger fragment than the sugar-phosphate-sugar we and the Clementi group used; and that we have a trial potential energy function with a parameter set of such quality that it is worthwhile to use it as a basis for further development.

6.3   Spiro compounds

In an attempt to see how our former "standard" potential energy functions PEF300 and PEF400, those used most extensively in our cal-culations, would behave towards the type of strain encountered with small rings, we studied a series of spiro compounds[221].

The parameter sets were modified slightly, as described in detail in the paper. This was done partly to conform to our new forms of potential energy functions, see sections 9.1, 9.2 and 11.6.2, partly to take into account the special problems encountered with torsional angles in spiro compounds containing small rings. The parameter sets were checked on cyclohexane, cyclopentane, cyclobutane and cyclopro-pane with good results except for the vibrational spectrum of cyclopropane and the structure of cyclobutane which came out planar as in some of its derivatives.

This point illustrates one conclusion of the work, that our potential energy functions are too primitive to account for large deviations from "ideal" geometry. Another conclusion was that our potential energy functions, despite their simple form and few parameters, are good enough for a first estimate of structure and thermodynamic functions even of these compounds.

6.4   Chloroalkanes

It is mentioned in chapter 8 that we have collected experimental data for about a hundred conformers of chloroalkanes. They will be used in a large-scale optimisation employing also the program MOL-VIB[253]. Some exploratory work was done before[254]: with a simple harmonic potential energy function we calculated the conformation and vibrational spectra of 2-chloropropane and two deuteromers. We also demonstrated the value of performing energy minimisation prior to normal coordinate analysis.

It is the intention to proceed with this work when the new optimisations on alkanes (see sections 9.1, 9.2, and 11.6.2) have been perfected.

7   Optimisation: Algorithms and implementation

L´humilité est pareille à un manteau;
parfois il faut savoir l´enlever.

Le grand Maguid de Mezeritch

Some readers have complained that chapter 7 of ref. 190 is too lapidaric and difficult to understand. As there are also too many errors in it (mine, not the typist´s), and as I have changed the optimisation algorithm and widened the scope of optimisation, it seems necessary to rewrite that chapter completely.

7.1   The optimisation algorithm

We revert to the original formulation of ref. 151. It was not realised at the time that the original algorithm is really a Levenberg – Marquard one; compare with for instance ref. 180.

Let $\underline{p}$ be a vector whose components are the current values of those potential energy function parameters we want to optimise:

$$\underline{p} = |p_1 \ p_2 \cdots p_m \cdots p_{nopt}\rangle;$$

nopt is the number of parameters to be optimised.

$\underline{\delta p}_m$ shall be a vector whose only non-zero component is δp_m, a small change in the value of p_m:

$$\underline{\delta p}_m = |00 \cdots \delta p_m \cdots 0\rangle.$$

A general change in the current value of $\underline{p}$ may now be written

$$\underline{\delta p} = \sum_{m=1}^{nopt} \underline{\delta p}_m.$$

Let y be a vector whose components are the values of those individual observables we select to optimise on:

$$\underline{y}^{calc} = |y_1 y_2 \ldots y_k \ldots y_{ntot}\rangle^{calc},$$
$$\underline{y}^{meas} = |y_1 y_2 \ldots y_k \ldots y_{ntot}\rangle^{meas};$$

ntot is the total number of observables to optimise on, of whatever
kind they may be; the count runs over all molecules in the set, and
for each molecule over bond lengths, valence angles, torsional
angles, vibrational frequencies, dipole moment etc.

The difference between a calculated value and the corresponding
measured value of an observable is

$$\Delta y_k = y_k^{calc} - y_k^{meas}.$$

From these we form a difference vector

$$\underline{\Delta y} = |\Delta y_1 \Delta y_2 \ldots \Delta y_k \ldots \Delta y_{ntot}\rangle.$$

The problem is now: Given $\underline{p}$, find a $\underline{\delta p}$ to make $\underline{\Delta y}$ smallest possible.

The solution is to minimise the quadratic norm of the weighted
$\underline{\Delta y}$:

$$(\Delta y´W)(W\Delta y)$$

where W is a diagonal weighting matrix, and where we have dropped
the vectorial notation, which we shall do from now on.

This is a least-squares problem, and a very much non-linear
one, as Δy is an extremely complicated function of p. We therefore
proceed to linearise the problem by expanding Δy around p:

$$\Delta y(p+\delta p) = \Delta y(p) + Z\delta p + \ldots$$

Z is the Jacobian matrix of partial derivatives of observables with
respect to potential energy function parameters:

$$Z_{km} = \frac{\partial \Delta y_k}{\partial p_m} = \frac{\partial(y_k^{calc} - y_k^{meas})}{\partial p_m} = \frac{\partial y_k^{calc}}{\partial p_m}.$$

To justify neglecting second and higher order terms in the expansion, we must keep δp small and hope for neat behaviour of $y(p)$.

Many weighting schemes have been proposed. As a standard, our consistent force field program uses the same as Lifson and Warshel[151]: W_{kk} is the reciprocal of the absolute uncertainty in y_k^{meas}.

With the residual

$$R = \Delta y(p+\delta p) = \Delta y(p)+Z\delta p$$

the weighted squared residual becomes

$$
\begin{aligned}
S &= \Delta y´(p+\delta p)WW\Delta y(p+\delta p) \\
&= (\Delta y(p)+Z\delta p)´WW(\Delta y(p)+Z\delta p) \\
&= \Delta y´(p)WW\Delta y(p)+\Delta y´(p)WWZ\delta p \\
&\quad +\delta p´Z´WW\Delta y(p)+\delta p´Z´WWZ\delta p.
\end{aligned}
$$

We now seek δp as a solution to a set of nopt equations

$$\frac{\partial S}{\partial \delta p´} = Z´WW\Delta y(p)+Z´WWZ\delta p = 0.$$

Here we must put in a constraint because of the need to keep δp small:

$$Z´WW\Delta y(p)+Z´WWZ\delta p+\lambda\delta p = 0,$$

where λ is some scalar, and we obtain the solution as

$$\delta p = -(Z´WWZ+\lambda I)^{-1}Z´WW\Delta y(p)$$

where I is the identity matrix.

This is done via Cholesky decomposition of the modified Z matrix exactly as described in ref. 190, chapter 5, using the same subroutines.

When δp is found, a fast estimate can predict the decrease in S provided linearity holds:

$$
\begin{aligned}
S(p+\delta p)-S(p) &= \Delta y´(p)WWZ\delta p \\
&\quad +\delta p´Z´WW\Delta y(p)+\delta p´Z´WWZ\delta p \\
&= \Delta y´(p)WWZ\delta p+\delta p´\delta p
\end{aligned}
$$

The choice of λ is difficult. Here we have chosen an initial value of $1000\overline{\text{Vnopt}}$ which fits with a stop criterion of $\delta p'\delta p < 10^{-6}$.

In iterations following the first, λ is modified according to the change in S: if S decreases, λ is decreased, if not, λ is increased.

It is also easy to predict Δy of the following iteration:

$W\Delta y^{new} = W\Delta y^{old} + WZ\delta p$,

but that is probably not of great interest.

7.2   Termination criteria

The following criteria were chosen:

(1)   Number of iterations equals a value set in the input: icycl = noptim.

(2)   Parameter changes very small: $\delta p'\delta p < 10^{-6}$.

(3)   Gradient of S very small: $\|\nabla S\|^2 < 10^{-5}$.

(4)   Relative changes in S and p small:

$$S_{k-1}-S_k < 0.005\ S_{k-1} \text{ and } \delta p/p < 0.001$$

(5)   Divergence:

$$S_{k-1}-S_k > 0 \text{ and } |(S_{k-1}-S_k)/S_k| > 0.1$$

Stopping because criteria (2-4) are met is considered desirable and is in consequence called ´optimal´ in the output. Stop by criterion (5) is called ´fiasco´, and the actual value of λ is printed, with a proposal to increase it .

7.3   The partial derivatives

The elements of the Jacobian or Z matrix are the partial derivatives of observables with respect to potential energy function
parameters. Some classes of them are rather laborious to obtain;
this is the price we have to pay for being able to optimise simultaneously on many classes of observables.

At the time of writing we can optimise on five classes of observables: geometry expressed as the internal coordinates bond lengths,
valence angles and torsional angles; rotational constants; atomic
charges; dipole moment; and internal frequencies of vibration.

The choice of internal coordinates as an object for optimisation
is obvious; use of rotational constants maybe less so. They certainly do not give very detailed information about the conformation
of a molecule, but they are the primary structural information
derived from rotational and ro-vib spectroscopy on small molecules.
The inclusion of dipole moments is a must when Coulomb terms are
present in the potential energy function. Charges are included, although they are not experimentally observable quantities, because it
may be desirable to lock a parameter set to data derived from photoelectron spectroscopy or from ab initio calculations with a large
basis set. Quite naturally we want to optimise on vibrational spectra, and we shall see below that it is a bit more cumbersome in the
consistent force field context than in traditional normal coordinate
analysis.

As far as I know, optimisation on rotational constants, charges
and dipole moment is new in the consistent force field context.

It might be desirable to optimise also on other quantities, for
instance thermodynamic properties. Until now, we have not done this,
for several reasons. One is, to be honest, that I have little lust
for spending a larger part of my life on programming. Another that
the approximations behind the statistical summations used by most
people and also by us [100,216] become inaccurate for open-chain molecules already from about butane. Further, the most interesting

thermodynamic data from a conformational analysis point of view, whether for simple molecules, saccharides or coordination complexes, are differences between thermodynamic functions of different molecules. Calculation and intermediate storage of such quantities is easy, computation of the Z-elements is not. I have not found a solution to the problem of one Z element belonging to different molecules; neither has anyone else, as far as I know. Lifson and Warshel [151] restrict, if my interpretation is correct, the deviation in a difference between two molecules to a deviation in one of them. Finally, the following point of view should be considered: There should be ´something´ left out from optimisation, to see whether a final potential energy function is not only ´internally´ consistent, but also consistent with ´external´ data. The argument came from Martin Karplus during our many discussions in Rehovot in the last months of 1969. I should add that it did not then coincide with the views of Shneior Lifson, but has been used in recent work[154].

$$\underline{7.3.1} \quad \underline{Internal\ coordinates}$$

The elements of Z are

$$Z_{km} = \frac{\partial y_k}{\partial p_m} = \sum_{1} \frac{\partial y_k}{\partial x_1} \frac{\partial x_1}{\partial p_m}$$

or

$$Z = B \frac{\partial x}{\partial p} ,$$

where the B matrix is the same as that used in section 6.2^{190}.

We now have to find the derivatives in cartesian coordinates x

$$\partial x/\partial p_m = |\partial x_1/\partial p_m, \partial x_2/\partial p_m, \ldots, \partial x_1/\partial p_m, \ldots\rangle .$$

They are defined as

$$\frac{\partial x_o(p)}{\delta p_m} = \lim_{\delta p_m \to 0} \frac{x_o(p+\delta p)-x_o(p)}{\delta p_m}$$

where the subscript $_o$ denotes equilibrium conformation.

In principle, x_o is a known function of p and therefore also of δp_m, through the equilibrium conditions

$$\nabla V(x_o(p);p) = 0$$

$$\nabla V(x_o(p+\delta p_m);p+\delta p_m) = 0.$$

In Chapter 5 of ref. 190, $x_o(p)$ is found from the Taylor expansion

$$\nabla V(x(p);p) = \nabla V(x_o(p);p)+F(x_o(p);p)\delta x$$

and the above condition as

$$x_o(p) = x(p)-F^{-1}(x_o(p);p)\nabla V(x(p);p)$$

where x is an arbitrary initial conformation.

Note that, in stead of the original F, which is singular or nearly singular, we use the modified F^{190}.

$F(x_o(p);p)$ is approximated with $F(x(p);p)$, and the equation is solved by iteration.

Analogously, $x_o(p+\delta p_m)$ may be found from

$$\nabla V(x(p+\delta p_m);\delta p_m) = \nabla V(x_o(p+\delta p_m);\delta p_m)+F(x_o(p+\delta p_m);\delta p_m)\delta x$$

and the equilibrium condition as

$$x_o(p+\delta p_m) = x(p+\delta p_m)-F^{-1}(x_o(p+\delta p_m);\delta p_m)\nabla V(x(p+\delta p_m);\delta p_m).$$

As the arbitrary initial conformation x we may choose the conformation we know from minimisation, $x_o(p)$:

$$x_o(p+\delta p_m) = x_o(p)-F^{-1}(x_o(p+\delta p_m);\delta p_m)\nabla V(x_o(p);\delta p_m).$$

Approximating F computed at equilibrium with F at the initial conformation, just as before, we get

$$x_o(p+\delta p_m) = x_o(p) - F^{-1}(x_o(p);\delta p_m) vV(x_o(p);\delta p_m)$$

and, as a difference quotient,

$$\frac{\partial x_o(p)}{\partial p_m} = \frac{x_o(p+\delta p_m) - x_o(p)}{\delta p_m} = -F^{-1}(x_o(p);\delta p_m)\nabla V(x_o(p);\delta p_m)/\delta p_m.$$

This means that both the gradient and the Hessian matrix are calculated at the equilibrium conformation as found with unchanged energy parameters, but now with one parameter changed at a time. The set of linear equations can be solved by standard methods.

Lifson and Warshel[151] used another approximation for F, $F(x_o(p))$. Therefore they computed the Hessian matrix only once per molecule per iteration step in the optimisation, whereas we do it nopt times per molecule per iteration. However, the programme MOLEC and all its subprogrammes must anyway be called because the gradient is needed, and our algorithm avoids many transports of the large Hessian from background memory. In addition, I have checked numerically that this method is more accurate.

7.3.2 Rotational constants

For some years, the consistent force field program has had an option for transformation of molecular coordinates to a coordinate system defined by the principal moments of inertia of the molecule. Once this has been done, the moments are found as

$$I_A = \sum_{i=1}^{natom} m_i(y_i^2 + z_i^2) \text{ etc.,}$$

and the derivatives in cartesian coordinates as

$$\frac{\partial I_A}{\partial x_i} = 0, \quad \frac{\partial I_A}{\partial y_i} = 2m_i y_i, \quad \frac{\partial I_A}{\partial z_i} = 2m_i z_i \text{ etc.}$$

As

$$\frac{\partial I_\alpha}{\partial p_m} = \sum_{j=1}^{3*natom} \frac{\partial I_\alpha}{\partial x_j} \frac{\partial x_j}{\partial p_m}, \quad \alpha = x,y,z,$$

and, as $A = const/I_A$ etc., we get

$$\frac{\partial A}{\partial p_m} = \frac{const}{I_A^2} \sum_{k=1}^{natom} \frac{\partial x(k*3-2)}{\partial p_m} *2m_k(x(k*3-1)+x(k*3))$$

and correspondingly for B and C.

const is = 505.391, if A is in GHz, m in dalton, and cartesian coordinates in Ångstrøm.

7.3.3 Atomic charges

The partial derivatives are very easily gotten as

$$\frac{\partial e_i}{\partial p_m} = \frac{e_i(p+\delta p_m)-e_i(p)}{\delta p_m}$$

7.3.4 Dipole moments

Here we have two possibilities:

$$\frac{\partial \mu}{\partial p_m} = \frac{\mu(p+\delta p_m)-\mu(p)}{\delta p_m}$$

and, as

$$\mu = \sum_{i=1}^{natom} e_i x_{i\alpha} \,, \quad \mu = (\mu_x^2+\mu_y^2+\mu_z^2)^{1/2} = (\sum_\alpha u_\alpha^2)^{1/2}$$

$$\alpha = x,y,z,$$

$$\frac{\partial \mu_\alpha}{\partial p_m} = \sum_{i=1}^{natom} (e_i \frac{\partial x_i}{\partial p_m} + \frac{\partial e_i}{\partial p_m} x_{i\alpha})$$

$$\frac{\partial \mu}{\partial p_m} = \frac{1}{\mu} \sum_\alpha \mu_\alpha \frac{\partial u_\alpha}{\partial p_m}$$

$$\alpha = x,y,z,$$

Both formulations have been tested; they give the same results, apart from tiny differences due to round-off.

7.3.5 Internal frequencies

The frequency subset of the Z matrix is found through the eigen-values of vibration

$$\lambda_k = (\text{const} * y_k)^2$$

as

$$\frac{\partial y_k}{\partial p_m} = \frac{\partial y_k}{\partial \lambda_k}\frac{\partial \lambda_k}{\partial p_m} = \frac{1}{2(\text{const})^2 y_k}\frac{\partial \lambda_k}{\partial p_m}.$$

In the original formulation[151,190], the following algorithm was used.

$\partial \lambda_k/\partial p_m$ are found from the secular equation as

$$\partial \lambda_k/\partial p_m = \delta q_k^{'}(\partial \lambda/\partial p_m)\delta q_k = \delta q_k^{'}L^{'}(\partial H/\partial p_m)L\delta q_k$$

$$= \delta q_k^{'}L^{'}M^{-1/2}(\partial F/\partial p_m)M^{-1/2}L\delta q_k.$$

The derivative of the Hessian is, by definition,

$$\frac{\partial F}{\partial p_m} = \lim_{\delta p_m \to 0} \frac{F(x_o(p+\delta p_m);\delta p_m)-F(x_o(p);p)}{\delta p_m} \approx \frac{F(x_o(p);\delta p_m)-F(x_o(p);p)}{\delta p_m}$$

The first term in the nominator is the Hessian calculated at the equilibrium conformation corresponding to a changed value of parameter p_m. An approximation to this equilibrium conformation is known from section 7.3.1, and the Hessian is found by a call of MOLEC. The second term is the equilibrium Hessian for unchanged parameters.

The L matrix is read from background memory, where it was written by VIBRAT, and the equilibrium Hessian from another background file, where it was written by CONFOR.

In the second, approximate, formulation, the first term is the Hessian calculated with changed parameter m, but at the equilibrium conformation for unchanged parameters. This is the formulation which was originally used by Lifson and Warshel[151] but probably not stuck to. In recent tests it gave unreliable derivatives.

Recently a new formulation has been used, which avoids the need for transporting masses of eigenvectors.

The Hessian $F(x_o(p+\delta p_m); \delta p_m)$ is calculated at the conformation $x_o(p+\delta p_m)$ which is available, and $F(x_o(p); p)$ is read from background and subtracted.

The difference is divided by δp_m and mass-weighted, and the eigenvalues are found directly by diagonalisation, not by transformation with the diagonalising L matrix as found at equilibrium. This gives directly $\partial\lambda/\partial p_m$ and is more accurate. The change is costlier in computing and cheaper in data transport. On a big machine like IBM 3081 this is costlier, but may be favourable on others. The procedure has in my tests given results comparable, but not identical, to those of the first formulation.

A different algorithm, proposed by Dr. C. J. M. Huige, Utrecht, is the following.

It calculates directly the frequency derivatives

$$\frac{\partial y_k}{\partial p_m} \approx \frac{y_k(x_o(p+\delta p_m); \delta p_m) - y_k(x_o(p); p)}{\delta p_m}.$$

The first term in the nominator is found as $(const)\sqrt{\lambda_k}$, with λ_k calculated in the same way as $\partial\lambda_k/\partial p_m$ in the original formulation, using the equilibrium L matrix. Transfer of the equilibrium Hessian is therefore avoided. The second term is the normal frequency known from VIBRAT.

Test runs give results only marginally different from those of the first algorithm, which is, of course, a check of the correctness of addressing vectors and matrices. It is only marginally cheaper than the former.

In a variant of this, $y_k(x_o(p+\delta p_m); \delta p_m)$ is calculated by mass-weighting of $F(x_o(p+\delta p_m); \delta p_m)$ and diagonalisation. This procedure is the most input-output saving, as also transfer of the L matrix is avoided. Because of the diagonalisation it is costlier. The results, in our tests, are almost indistinguishable from those of the first algorithm, but the Z matrix elements are more precise, judging from the observation that derivatives which should be zero are below 10^{-12}, which rarely happens with the first algorithm.

This last is by far the most transparent and gives the simplest code.

7.4 Correlation and uncertainty

The interdependence of the potential energy function parameters cannot be judged by inspection of the correlation coefficients as is usual in optimisation because of the methods employed here: the modified Z matrix is not diagonalised. For the same reason, an estimate of the standard deviation is not possible either.

7.5 Organisation of the optimisation

Reading of experimental data is done with a subroutine RDEXP, which is called once per job. It runs over the molecules, and for each molecule it runs over the types of observables. The following data are read: The list number of each internal coordinate, rotational constant, atomic charge, dipole moment (always 1) and internal frequency on which optimisation is desired; its experimental value; its experimental uncertainty. The uncertainty may be left out; if so, it is given a standard value. This is practical for frequencies.

The routine counts the number of data of each type for each molecule and the total number of data. All the information is stored in a background file.

The programme OPTIM controls the optimisation. Through calls of a subroutine NPAR it changes by small amounts the values of those parameters that are to be optimised, one by one, and through subroutine BUILDZ the elements of the Z matrix are built up, one row per parameter to be optimised. The rows are written on a background file. Each element is multiplied by the proper element of the weighting matrix.

The Z matrix is put together from its rows by a subroutine ZMATRX, and is printed for inspection, one set of lines of elements per parameter to be optimised.

The WΔy vector is constructed by a subroutine BUILDY, which has access to the background file of experimental data and to files of calculated observables prepared during the current optimisation cycle.

OPTIM performs the necessary matrix algebra on the outputs from BUILDZ and BUILDY, and finds δp by call of CHLSKY and LINSOL (ref. 190, chapter 5). It also administers the termination criteria, and updates the parameters.

The interrelation of programs, their hierarchy, and all data transports between them are symbolised in Figure 7-1.

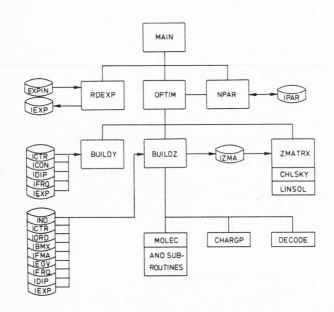

Figure 7-1: Subprograms used in optimisation

8   A data bank for optimisation

By Kjeld Rasmussen and Birgit Rasmussen

It is easier to ponder relationships
between meaningful numbers
than to gather them

Engler, Andose and Schleyer[70]

The required format for input of experimental data to be used in optimisation is described in the input manual, see A2. In order to ease the compounding of optimisation input we have built a bank of molecular data.

The data bank presently has the following departments: structure, frequency, charge, dipole moment; more will be added as need arises, notably rotational constants, lattice energy and unit cell dimensions.

Within a department there is an account for each molecule for which data have been selected; for each account there may be any number of records. A record consists of: list number of the internal coordinate or frequency, its type, its value, its uncertainty. The last is usually chosen as 3*(standard deviation).

The data bank presently holds records for the following classes of substances: alkanes, cycloalkanes, chain and cyclic ethers, alcohols, chain and cyclic esters (lactones), chain and cylic amides (lactams), amines and ammonium ions, saccharides, coordination complexes, chloroalkanes. More classes and more substances are being added as optimisation progresses.

8.1   Structure

The most desirable structure to optimise on, when the objective
is to develop potential energy function parameters for calculation
of conformation by energy minimisation, is the equilibrium structure
r_e. This structure is known for so few compounds that its use in
this connection is out of question. Especially so, since a set of r_e
cannot be supplemented with a set of other structures, as an r_e is
far "smaller" than any other structure.

Molecular structures derived from electron diffraction and microwave spectroscopy are summarised and exemplified by Kuchitsu[140], who
also describes their determination.

The thermal-average structure r_g is that usually reported by
electron diffractionists, and it is the most suitable representation
if comparisons only of bond lengths in closely related molecules are
wanted.[141,142] It is not so suitable for angles, and for different
classes of molecules. It is also wrong in principle to use tempera-
ture-dependent dynamically-averaged data to fit a model which is
static in concept and realisation.

We chose, therefore, to optimise on the structure r_α^o or r_z which
is the structure given by the average nuclear positions in the
vibrational groundstate at 0 K, derived from electron diffraction or
microwave spectroscopy or, preferably, both.[140,141] Quite many r_α^o
and r_z structures have been published, and where only the r_g is
reported from an ED experiment, it is often possible to obtain the
r_α^o from the author. When this is not the case, one may have to compromise with other average structures. Valence angles are the same
in r_α^o and r_α, and often in r_g as well, to within 3σ or better,
though here one must be careful, especially in cases of large-ampli-
tude low-frequency motion[141].

As an example and, hopefully, a guidance, we summarise here our
selection of the first entries in the structure department. Table
8-1 shows a few examples of records extracted from the conformation
department of the bank.

Table 8-1: Ethane, n-butane and cyclohexane

1CC	1.5323	0.006				
2CH	1.1017	0.006	19HCH	1.8764	0.016	101
39CCCC	+1.13	0.10				106
1CC	1.532	0.006				
2CH	1.104	0.012	21CCC	1.944	0.011	
25HCH	1.876	0.079	56CCCC	0.958	0.021	302

8.1.1   Alkanes

For ethane, the $r^o_\alpha = r_z$ is known to very high precision[118]. For propane, also the $r^o_\alpha = r_z$ is known[117]. The r_s is significantly different[150]. For isobutane, the $r^o_\alpha = r_z$[108] is a very detailed structure. For neopentane, only the C-C is of interest, and only the r_g is known[17]. For n-butane, the r_g is known[31]. For our purposes we use the torsional angle C-C-C-C as the only datum. Tris-(tert-butyl)-methane, an extremely crowded molecule, has been a test case[96,60] since its r_g structure[44] was published. It is our aim to use it for checking a parameter set rather than for optimising it, as it was done with an earlier potential energy function[217].

8.1.2   Cycloalkanes

For cyclohexane, the definitive r_α is used[18]. For cyclopentane we included the r_a derived directly from experiment[4].

8.1.3 Ethers

For dimethylether, all structures r_s, r_g, even r_e, are known, except r_z. We use the r_a[133]. For methylethylether we have made a comparison between the r_s[106] and the r_α^o[199,200]; we use the latter in optimisation. To broaden the comparison of structures, we include the r_s of methylpropylether[129] and diethylether[105] in the bank.

The anomeric CO bond presents a special problem in carbohydrate structure, which is a main activity of our group[215], and we have therefore included in the data bank a number of model molecules in which a tetrahedral carbon atom is bound to two or more oxygen atoms. For bis-(methoxy)-methane[11] and bis- (methoxy)-propane[12] the r_a are known; for tetrakis- (methoxy)-methane the r_g[173].

8.1.4 Cyclic ethers

Tetrahydrofuran is included though, as for cyclopentane, only the r_a is known[5], because it is central as a model for furanosides. For tetrahydropyran, a very detailed and accurate r_α^o is known[33,90]. An old r_a is known for dioxan[57]. Paraformaldehyde has exclusively ano-meric CO bonds. Many structures are published, among them high- and low-temperature x-ray[46,47] r_o from MW[196], and r_g[48]. The latter is used in our comparison. Paraldehyde contains the same skeleton, plus C–C bonds; the r_a is known[10].

8.1.5   Alcohols

For methanol, many, and interesting, structures are known. A selection: r_a[133], r_o[146,149], very close to r_a, and structures derived from x-ray and neutron diffraction in the liquid; except for O-H (or rather O-D) very close to r_o[158,178]. For ethanol, many measurements of A_0, B_0, and C_0 of both conformers, protonated and deuterated, are found[257,233,128,127]. For optimisation we use a structure derived from neutron diffraction in the liquid[179]. For propanol-2, only A_0, B_0 and C_0 of both conformers are known[138]. For propanol-1, A_0, B_0 and C_0 are known for the conformers gauche-gauche and anti-gauche[2,1].

8.1.6   Esters and lactones

A recent r_s of methylformate has been published[19]; also A_0, B_0, C_0 are given. An ancient ED investigation of methylacetate[195] is not of much use in this context; the torsion around the -C-O- bond has been reported later[251] and is included for optimisation. Lactones are important to us for checking purpose; unfortunately no detailed structure has been published. A_0, B_0 and C_0 have been measured for γ-butyrolacton[66], γ-valerolacton[131] and two conformers of δ-valerolacton[205].

It is relevant to refer to a recent review of crystal structure of esters by Schweizer and Dunitz[238], which may be useful also in the present context.

8.1.7   Amines

For the purpose of optimising a potential energy function for transition metal complexes with coordinated diamines, another main activity of our group[216], we have selected a few diamines and diammonium ions for the data bank.

A few conformer structures of 1,2-ethanediamine are known: the r_g of the gauche conformer[287]; the r_o of two gauche conformers[160] having different N-C-C angles; a low- temperature x-ray structure of the anti conformer[123]; and an x-ray structure of the bis-borane complex of the anti conformer[258].

Many x-ray structures are available of the 1,2-ethanediammonium ion. We have selected two different structures of the anti conformer having the same quality[61,22], and a very accurate one of the gauche conformer[79]. Also a neutron diffraction structure of the anti conformer is used[75].

1,3-diaminopropane has such a complicated and weak microwave spectrum that we shall probably never obtain rotational constants and dipole moments[175]. An x-ray structure of its diammonium ion is available[110].

8.1.8   Chloroalkanes

75 conformers of chloroalkanes from ethanes to butanes are also included, almost all with their r_a structure. They are not described further here, but will be discussed in connection with the optimisation of a potential energy function for chloroalkanes.

8.2   Vibrational frequency

Empirically well-assigned spectra are taken from Shimanouchi´s compilation[243] if possible, and otherwise from original articles.

In our selection of material we have used the following guidelines: C-H, H-C-H, C-C-C and similar deformations are taken from the simpler compounds, ethane, propane, cyclohexane; deformations involving heteroatoms from progressively more complicated substances; torsional deformations from wherever they have been observed and assigned.

8.3   Fractional atomic charge

This department is the smallest of the bank. It is used mainly to limit variation of charge parameters by optimisation on atomic charges within a few selected molecules, with "experimental" values mostly from Mulliken population analysis of results from ab initio calculations with reasonably large basis sets, and using small "uncertainties". Also data from photoelectron spectroscopy are occasionally used.

8.4   Dipole moment

Dipole moments are taken from microwave measurements where possible, else from solution studies.

8.5 Rotational constants

Rotational constants are to be used in optimisation only where this is the only structural information available, for instance where only a microwave study of a single or too few isotopomers has been made.

8.6 Unit cell dimensions

In optimisation on crystal properties, the first observables used were the unit cell dimensions, one to six data, which for technical reasons are concatenated to the conformational observables. Next, lattice energies were introduced, so far in a limited number. The use of other properties is being considered.

8.7 Statistical test

When we had selected model compounds and structures on which to optimise, we felt the need for a test of their consistency. We chose to use the Statistical Analysis System (SAS) for comparing individual and average values of bond lengths and angles, and standard deviations and ranges, for different classes of molecules. The goal was to indicate whether or not the same potential energy function parameter values can be used for, say a C-C bond in an alkane, a cyclic ether, and an ester; and to be guided in the selection of individual data for the optimisation.

A SAS program was written which will read directly from a concatenation of data bank records, perform a simple analysis of the material in terms of mean values and deviations of each type of observable, and plot the results on the line printer.

Figure 8-1 shows the merging of several such observations; we shall comment shortly on it, just to show how we go about selecting data for optimisation with SAS, once the initial selection for the data bank has been done.

Bond lengths are given in Å and angles in rad. For average values, standard deviations are shown; 1.099 (6) means that the average is 1.099 Å with a standard deviation of 0.006 Å. This gives no information about experimental accuracy; for individual values we show the observed datum with three times the reported standard deviation of the measurement. In ethanol, C-O-H 1.873 (21) means that the C-O-H angle is 1.873 rad with a standard deviation on the measurement of 0.007 rad.

8.7.1 Alkanes and cycloalkanes

The average C-C bond is 1.536 (4) whether or not the two cycloalkanes are included. This is so because the r_a of cyclopentane enters with a low weight; it is significantly larger. For the three alkane r_z alone the average is 1.532 (1), which is identical to r_a for cyclohexane. The conclusion is that all data may be used in the same optimisation, but that neopentane and cyclopentane may be excluded without loss of essential information.

Completely analogous analyses were made for C-H bonds, and C-C-C, C-C-H and H-C-H angles.

The two C-C-C-C torsions, in n-butane g and cyclohexane, cannot be compared in this way. They are very important in optimisation especially for the determination of non-bonded potential energy function parameters.

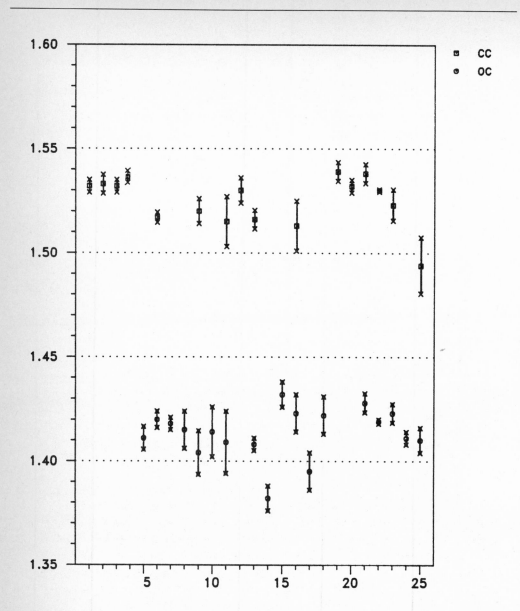

Individual entries are:

1   r_z ethane
2   r_z propane
3   r_z isobutane
4   r_g neopentane
5   r_z methylethylether, methyl end
6   r_z methylethylether, ethyl end
7   r_a dimethylether
8   r_s methylethylether, methyl end
9   r_s methylethylether, ethyl end
10 r_s methylpropylether, methyl end
11 r_s methylpropylether, propyl mid
12 r_s methylpropylether, propyl end
13 r_s diethylether
14 r_a bis-methoxy-methane, central
15 r_a bis-methoxy-methane, peripheral
16 r_a bis-methoxy-propane
17 r_g tetrakis-methoxy-methane, peripheral
18 r_g tetrakis-methoxy-methane, central
19 r_a cyclopentane
20 r_α cyclohexane
21 r_a tetrahydrofuran
22 r_z tetrahydropyran
23 r_a dioxane
24 r_g trioxane
25 r_a paraldehyde

Figure 8-1: C-C and O-C bond lengths with +3σ intervals

8.7.2   Ethers and cyclic ethers

The average value of C–C is 1.524 (8), when all 7 non-anomeric substances are treated. If we leave out methylpropylether, which has a large spread, and tetrahydrofuran, an r_a, the average is 1.521 (6), which is not significantly different. The uncertainty intervals of r_s for diethylether and of r_α^o for methylethylether are almost identical; of r_a for dioxan wider, but overlapping. r_α^o for tetrahydropyran is much narrower, and at a larger value. The C–C bond is anyway shorter than in the alkanes, by about 0.010.

The O–C bond has the average value 1.411 (5) in the r_s of the straight-chain ethers and 1.423 (4) in the r_a of the cyclic ethers; r_α^o of tetrahydropyran is 1.419 (2). From a plot of the r_s we find that –O–CH$_2$ is constant, that –O–CH$_3$ is significantly longer, and that –O–CH$_3$ decreases with increasing chain length. This tendency might presumably apply to any structure type, but such is not the case. In the r_α^o of methylethylether the inequality is reversed: –O–CH$_3$ is 1.411 (11) and –O–CH$_2$ is 1.420 (8). Also for this reason r_s structures will not be used in optimisation.

The special problems encountered in molecules where two or more oxygen atoms are bound to one tetrahedral carbon atom are illustrated in the following.

C–C is 1.513 (24) in bis-(methoxy)propane, but 1.494 (27) in paraldehyde, both from the r_a. Both are significantly shorter than in normal straight-chain and cyclic ethers.

The average O–C is 1.411 (17), which is not significantly different from the other ethers. Individual bond lengths are significantly different in two of the branched ethers, the central being 1.382 (12) and the peripheral 1.432 (12) in bis-(methoxy)-methane. It is rather as if the difference has been exaggerated in the structure fitting process; in bis-(methoxy)-propane both are given as 1.423 (18), and in tetrakis-(methoxy)-methane the two are given as 1.395 (18) and 1.422 (180).

A similar analysis of other properties than structural data is not envisaged.

8.8   Use of the data

A few notes on how the selected data are used in optimisation are relevant. As an example, let us look at the C-C bond in alkanes and cycloalkanes. The constancy of the C-C bond length does not mean that we have here a good value for the potential energy function parameter b_0. The C-C bond length in an actual molecule is in our model determined by other parameters as well, most clearly illustrated in ethane where H---H non-bonded interactions are equally important.

A conformer is the result of a delicate balancing of forces. The C-C bond length in these compounds is therefore not an intrinsic property of the bond, at least in our model. It is therefore not important that it has the same value in ethane and cyclohexane; it is important that the rather different forces should balance so as to produce the same value.

If this is done properly, the C-C bond length in a much more strained molecule, as the overcrowded tris-(tert-butyl)-methane, will be given properly, without having been included in the optimisation. A previous potential energy function nearly did so[217].

If we then turn to ethers, we see that the C-C bond is shorter in the normal straight-chain and cyclic ethers than in alkanes. This must be taken care of in the optimisation of potential energy function parameters; different parameter values cannot be accepted, if the overall goal of a simple potential energy function for many classes of substances is to be attained.

The situation is different with the anomeric ethers. If we want a reproduction of structure, correct in all details, as a prerequisite for reliable predictions, we shall have to introduce one more type of atom, an anomeric carbon, with the necessary increased number of potential energy function parameters. Our analysis shows clearly that this is demanded by the C-C bond as much as by the C-O bond. For this purpose, the anomeric ethers are necessary in optimisation. In optimisation of parameters of normal carbon they must be excluded.

An old cloak makes a new jerkin;
a withered servingman a fresh tapster

The Merry Wives of Windsor: Act I, Scene III

Some of our first optimisations are reported here rather than in
the periodical literature. They supplement and improve our earlier
work in which parameter fitting was done by trial and error. A more
important aspect is that I try to pass on some of the know-how we
have gained, in order to assist future users of the system.

9.1   PEF300 series

The potential energy function with associated parameters called
PEF300 was developed[132] by trial and error. It gave quite nice
results for many properties of saccharides, see chapter 5, and the
alkane part of it was chosen first as a promising initial set of
parameters to be optimised.

9.1.1   Experimental data

From our data bank, see chapter 8, we selected the following data
for optimisation: The r_z structures of ethane[118], propane[117] and
isobutane[108], the torsional angle of n-butane gauche[31], and the r_α
structure of cyclohexane[18], plus selected low frequencies of ethane
(1), propane (3), isobutane (4), n-butane anti (5) and gauche (5),

Table 9-1: Parameters of the PEF300 series[a]

	PEF300		PEF301opt	
Interaction	K_b	b_o	K_b	b_o
C-C	720.	1.52	563.	1.5184
C-H	720.	1.09	670.	1.0996
	K_θ	θ_o	K_θ	θ_o
C-C-C	143.9	T_d	143.89	T_d
C-C-H	93.5	T_d	93.5	T_d
H-C-H	74.8	T_d	74.8	T_d
	K_ϕ		K_ϕ	
-C-C-	0.005		0.0406	

	A	B	C	ε	r^*
C---C	$236. *10^3$	4.32	297.9	0.0010	3.4756
C---H	$31.37*10^3$	4.20	120.8	0.0873	3.1876
H---H	$6.59*10^3$	4.08	49.1	0.0473	2.9740

	PEF302opt		PEF303opt	
	K_b	b_o	K_b	b_o
C-C	563.08	1.5157	563.077	1.5157
C-H	670.	1.0991	670.	1.0990

	K_θ	θ_o	K_θ	θ_o
C–C–C	142.45	T_d	142.447	T_d
C–C–H	93.5	T_d	93.5	T_d
H–C–H	74.8	T_d	74.8	T_d

	K_ϕ		K_ϕ	
–C–C–	0.7286		1.2809	

	$\sqrt{A}$	$\sqrt{B}$	$\sqrt{A}$	$\sqrt{B}$
C––	559.26	19.723	559.123	18.865
H––	160.69	7.812	160.137	7.746

---

a) Units are such that energy comes in $kcalmol^{-1}$, distances in Å and angles in °. T_d means the tetrahedral angle $arcos(-1/3)$.

and cyclohexane (4); in a few cases more frequencies were used. Frequencies were mostly taken from Shimanouchi[243], for n-butane also from Durig and Compton[67]. Altogether 14 conformational and 22 – 35 vibrational data were used.

For checking an optimised parameter set independently of the optimisation we included the following supplemental data: the CC bond length from the r_g structure of neopentane[20], the r_g structure of tris-(tert-butyl)- methane[44], the r_a structure of cyclopentane[4]; the entire vibrational spectra of ethane and cyclohexane[243]; parts of spectra of others; thermodynamic data of the small alkanes[241] and the two cycloalkanes[144,165], as well as the free enthalpy difference of the two n-butane conformers[31] and the enthalpy difference of the two populated n-pentane conformers[104].

Table 9-2: Conformations in the PEF300 series

Internal	PEF300	PEF301opt	PEF302opt	PEF303opt	exp
Ethane					
C-C	1.5242	1.5291	1.5289	1.5289	1.5323
C-H	1.0908	1.1012	1.1011	1.1011	1.1017
H-C-H	109.14	108.83	108.68	108.68	107.51
Propane					
C-C	1.526	1.533	1.533	1.533	1.533
C-H3	1.091	1.101	1.101	1.101	1.097
C-H2	1.092	1.103	1.103	1.103	1.096
C-C-C	110.6	111.2	111.2	111.3	112.0
H-C-H	109.0	108.6	108.4	108.4	107.9
Isobutane					
C-C	1.529	1.538	1.537	1.538	1.532
C-Ht	1.092	1.105	1.106	1.106	1.109
C-H3	1.091	1.101	1.101	1.101	1.092
C-C-C	110.1	110.5	110.5	110.5	110.8
C-C-Ht	108.8	108.5	108.5	108.4	108.1
C-C-H	109.9	110.3	110.5	110.5	111.4
C-C-H	110.2	110.6	110.7	110.7	110.2
H-C-H	108.7	108.2	108.1	108.1	108.7
H-C-H	109.0	108.7	108.5	108.5	106.5
n-Butane g					
-C-C-	65.	58.	62.	63.	65.

Cyclohexane

C-C	1.528	1.531	1.533	1.533	1.532
C-H	1.092	1.103	1.103	1.103	1.104
C-C-C	110.5	110.6	110.7	110.6	111.4
H-C-H	108.5	107.7	107.4	107.4	107.5

Neopentane

C-C	1.531	1.543	1.543	1.543	1.537

Cyclohexane

-C-C-	57.3	57.1	56.9	57.0	54.9

Cyclopentane

C-C	1.528		1.530	1.530	1.539
C-H	1.098		1.104	1.104	1.095

TTBM

Cq-Ct	1.639	1.640	1.611
Cq-Cm	1.564	1.564	1.548
Cq-Ct-Cq	116.5	116.5	116.0
Ct-Cq-Cm	113.3	113.4	113.0
Cm-Cq-Cm	105.3	105.3	105.7
Cq-Ct-Ht	100.8	100.9	101.6
Cq-Cm-Hm	112.8	112.8	114.2
Δ(CmCqCtHt)	15.0	14.9	10.8
Δ(CtCqCmHm)	16.6	16.1	18.0

Table 9-3: Vibrational frequencies in the PEF300 series

Average deviations <o(meas)-o(calc)>

Substance	Internals	PEF300	PEF301opt	PEF302opt	PEF303opt
Ethane					
	CH	− 60	+ 36	+35	+ 20[a]
	CCH,HCH	− 30	− 37	−46	− 25
	CC	+ 6	+ 61	+52	+ 60	+ 52
	−CC−	+172	+ 98	+37	+ 30	+ 7
Propane					
	CC	− 14	+ 59		+ 53	
	CCC	− 69	−102		−104	−104
	−CC−	+123	+ 64		+ 23	0
Isobutane					
	CC	− 46	− 17		− 25	
	CCC	− 48	− 85		− 90	− 92
	−CC−	+129	+ 74		+ 44	+ 20
n-butane a					
	CCC	− 44	− 68		− 66	− 70
	−CC−	+ 81	+ 30		+ 10	− 31
n-butane g					
	CCC	− 61	− 86		− 95	− 96
	−CC−	+ 70	0		− 21	− 44

Cyclohexane

CH	-136	- 36	- 38	
CCH,HCH		+ 50	- 53	
CC	- 35	+ 3	- 10	
CCC	- 72	- 98	-107	-108
-CC-	- 36	- 9	- 29	- 39

----

a) C_2D_6

9.1.2 Initial experiences

PEF300 uses Buckingham exp-6 functions for non-bonded interactions. A variant of it, PEF301, which uses Lennard-Jones 12-6 functions in the (ε, r^*) form and gives comparable results[218], was used in the initial phase.

The first optimisations showed that the program is safe in the sense that it almost newer abends. It was also quickly learned that many iterations must be specified in a run, ideally enough to bring the parameter set to an optimum, and that optimisation should be done on large sets of data and with variation of all those parameters which have sizeable Z matrix elements for the data used.

The value of lambda, the Lagrange multiplier in optimisation, see chapter 7, is a good indicator of how well the optimisation has run: lambda should always decrease. If an optimisation ends with a high value of lambda, it has been initialised with a parameter set too far from optimum, and the result may well be an optimum only in the technical sense; a different initial parameter set will probably give a different and maybe better optimum.

Table 9-4: Thermodynamics[a] in the PEF300 series

Substance		PEF300	PEF301opt	PEF302opt	PEF303opt	exp.	note
Ethane	S	234.4	230.9	229.0	228.2	229.5	a
	C	51.3	50.8	50.4	50.1	52.8	
Propane	S	278.8	272.0	268.9	267.4	270.4	a
	C	70.4	69.5	69.0	68.7	73.4	
Isobutane	S	304.4	294.6	291.1	289.1	295.1	a
	C	92.3	90.5	89.8	89.4	97.3	
Neopentane	S			300.0	297.7	306.6	a
	C			111.8	111.2	121.5	
n-butane	S			298.7	296.7	311.0	a,f
	C			88.1	87.8	97.9	
n-pentane	S				327.9	350.3	a,f
	C				106.4	120.6	
Cyclohexane	S	297.7	294.3	292.2	291.5	298.2	b
	C	101.8	100.5	99.0	98.7	106.3	
Cyclopentane	S				277.8	292.9	c
n-butane	a:g			0.76:0.24	0.77:0.23	0.54:0.46	
G (g-a)		3.05	2.38	2.86	2.98	2.08	d
n-pentane							
H (ag-aa)					4.61	2.43	e

----

a) S: S_{300}^0/Jmol^{-1}K^{-1}; C: $C_{p,300}^0$/Jmol^{-1}K^{-1}; for butane and pentane weighted with the calculated conformer population; exp. data taken from a correlation, ref. 241. b) Ref. 144. c) Ref. 165. d) Ref. 31, uncertainty 0.92. e) Ref. 104, for the range 160-300 K. f) Calc. results averaged over the conformer population.

It has therefore little meaning, in problems of optimisation so complicated as these, to speek of THE optimum in an objective way. Common sense, insight and Fingerspitzgefühl are still needed also when one is assisted by these powerful computational methods.

For example, it is necessary to give K_b(CH) a good initial value; otherwise it will stay put.

There is clearly correlation between parameters of different types, such as non-bonded parameters, K_ϕ and K_θ(CCC), and some are rather badly determined, such as (ε, r^*) for C---C, K_θ(CCC) and K_b(CC).

9.1.3   Initial results

The best result we obtained for PEF301 with the data mentioned above gave a parameter set shown in Table 9-1; the resulting conformations are shown in Table 9-2, vibrations in Table 9-3, and thermodynamic functions in Table 9-4.

From the (ε, r^*) parameters of PEF301opt, (A,B) parameters were calculated, and $(\sqrt{A}, \sqrt{B})$ sets chosen for C-- and H-- non-bonded interactions; H-- from H---H and C-- from C--H, as ε(C---C) was badly determined. Substituting this set we got PEF302; the values changed very little on renewed optimisation. The results are shown in Tables 9-1, 9-2, 9-3 and 9-4.

9.1.4   Further results

The good predictive power in respect of conformation is quite
remarkable, see in particular the data for tris-(tert-butyl)- methane, TTBM, especially when we emphasize that only 12 variable
parameters are used in PEF302 and PEF303; all θ_o are locked at the
tetrahedral value arcos(-1/3). Comparison with PEF300 is not fair,
as that set was not optimised, but I wish to point out that PEF302,
with single-atom parameters in the non-bonded part, gives just as
good results as PEF301 with two-atom parameters. Analogous observations were made by Lifson and Warshel[151] and Williams[284,p.8].

PEF303 is the same as PEF302, but optimised also on crystal properties, see section 11.6.2. As far as internal properties are
concerned, results are only marginally different from those of
PEF302.

The reproduction of frequencies is of course not nearly so good;
this would require many more potential energy terms with associated
parameters. Care was taken to reproduce the spectra in the CH and
intermediate regions so well that average deviations were not unduly
high, and the lowest frequencies as accurate as possible. Table 9-3
shows that this goal was attained, with the exception of modes which
are predominantly CCC bending. This K_θ is simply very insensitive.

In a couple of instances, frequencies of deuterated substances
were calculated in PEF302. For ethane-d_6 the results are not significantly different, see Table 9-3. For isobutane-d_1 the C-D str was
(meas, calc) (2149, 2180).

For the slightly pathological case of cyclopentane it is interesting to see that these very simple potential energy functions give
a rather nice representation of the structure; see also chapter 10.

9.2   PEF400 series

The potential energy function with associated parameters called
PEF400 was developed[172] by trial and error as PEF300 was, only in a
more rational way, and on a much larger experimental material. Also
PEF400 gave very good results for many properties of saccharides,
see chapter 5.

It has been our wish for a long time to optimise PEF400, and we
have now done it on the same data as were used for the PEF300
series.

Originally, PEF400 was rather different from PEF300 in the non-
bonded part, using Lennard-Jones in stead of Buckingham functions,
and electrostatic monopole interactions. As PEF302 and PEF303 also
use Lennard-Jones functions, one should expect the parameters of the
two series to converge towards each other as optimisation progress,
particularly if Coulomb interactions are added to PEF303.

PEF400 was first optimised in unchanged form; b_o, K_o, ε, r^* and e
were allowed to change; K_b, K_θ and θ_o were kept. Next, PEF400opt was
changed to PEF401 as shown in Table 9-8, and this parameter set was
optimised in the same way. For these optimisations, dipole moments
of propane and isobutane were included.

The resulting parameter set is shown in Table 9-5; and the
resulting conformations in Table 9-6, vibrations in Table 9-7, and
thermodynamic functions in Table 9-9.

For comparison, a version of the PEF300 series using charge parameters was also optimised. PEF304 uses fixed charges, see section
11.6.2; the charge assignment algorithms are described elsewhere[171,215,223]. The parameters are also shown in Table 9-5;
results obtained with this set differ significantly from those gotten with PEF302 or PEF303; see, as examples, Tables 9-3 and 9-7.

Table 9-5: Parameters of the PEF400 series[a]

Interaction	PEF400 K_b	PEF400 b_o	PEF400opt K_b	PEF400opt b_o
C-C	510.	1.509	510.	1.5086
C-H	670.	1.093	670.	1.1003

	K_θ	θ_o	K_θ	θ_o
C-C-C	50.	T_d	50.	T_d
C-C-H	71.	T_d	71.	T_d
H-C-H	75.	T_d	75.	T_d

	K_ϕ		K_ϕ	
-C-C-	0.001		0.0137	

	ε	r^*	ε	r^*
C---C	0.1	3.5	0.0596	3.4827
C---H	0.1	3.15	0.1334	3.1556
H---H	0.3	2.75	0.3345	2.7796

	e		e	
C.	-0.050		-0.0340	
H.	0.125		0.0797	

	PEF401opt		PEF304opt	
	K_b	b_o	K_b	b_o
C–C	510.	1.5094	563.077	1.5134
C–H	670.	1.1004	670.	1.0866
	K_θ	θ_o	K_θ	θ_o
C–C–C	50.	T_d	142.447	T_d
C–C–H	71.	T_d	93.5	T_d
H–C–H	75.	T_d	74.8	T_d
	K_ϕ		K_ϕ	
–C–C–	0.0226		1.2809	
	$\sqrt{A}$	$\sqrt{B}$	$\sqrt{A}$	$\sqrt{B}$
C––	435.63	14.584	559.123	18.856
H––	266.74	17.566	160.139	7.705
	e		e	
C.	-0.0372		0.0	
H.	0.0422		0.14	

a) Units are such that energy comes in $kcalmol^{-1}$, distances in Å and angles in $^\circ$. T_d means the tetrahedral angle $\arccos(-1/3)$.

Table 9-6: Conformations in the PEF400 series

Internal	PEF400	PEF401opt	PEF304opt	exp.
Ethane				
C–C	1.527	1.5309	1.5329	1.5323
C–H	1.095	1.1033	1.0899	1.1017
H–C–H	108.3	108.11	108.34	107.51
Propane				
C–C	1.529	1.532	1.531	1.533
C–H3	1.095	1.102	1.089	1.097
C–H2	1.098	1.107	1.093	1.096
C–C–C	111.8	111.7	110.2	112.0
H–C–H	108.	108.1	108.2	107.9
Isobutane				
C–C	1.532	1.535	1.530	1.532
C–Ht	1.094	1.111	1.097	1.109
C–H3	1.101	1.102	1.088	1.092
C–C–C	110.6	110.5	109.8	110.8
C–C–Ht	108.4	108.4	109.1	108.1
C–C–H		111.2	110.6	111.4
C–C–H	110.7	110.6	110.9	110.2
H–C–H		107.9	108.0	108.7
H–C–H	108.2	108.0	108.4	106.5
n-Butane g				
–C–C–	58.	56.	65.	65.

Cyclohexane

C–C	1.530	1.532	1.532	1.532
C–H	1.098	1.105	1.091	1.104
C–C–C	110.9	110.8	110.2	111.4
H–C–H	107.1	106.7	107.2	107.5

Neopentane

C–C	1.539	1.538	1.529	1.537

Cyclohexane

–C–C–	56.4	56.7	58.0	54.9

Cyclopentane

C–C	1.528	1.531	1.527	1.539
C–H	1.098	1.107	1.093	1.095

TTBM

Cq–Ct	1.630	1.637	1.637	1.611
Cq–Cm	1.565	1.568	1.557	1.548
Cq–Ct–Cq	117.7	117.9	116.7	116.0
Ct–Cq–Cm	114.9	115.0	113.6	113.0
Cm–Cq–Cm	103.5	103.4	105.0	105.7
Cq–Ct–Ht	98.8	98.5	100.6	101.6
Cq–Cm–Hm	113.4	113.8	112.7	114.2
(CmCqCtHt)	15.1	15.0	15.0	10.8
(CtCqCmHm)	15.7	15.4	15.5	18.0

Table 9-7: Vibrational frequencies in PEF401 and PEF304

Average deviations ⟨σ(meas)-σ(calc)⟩

Substance	Internals	PEF401opt	PEF304opt
Ethane			
	CH		+ 34
	CCH,HCH		- 61
	CC	+ 38	+ 48
	-CC-	0	+ 6
Propane			
	CCC	- 28	-110
	-CC-	+ 4	7
Isobutane			
	CCC		- 20
	-CC-		+ 14
n-butane a			
	CCC	- 11	- 76
	-CC-	- 21	- 25
n-butane g			
	CCC	- 40	-103
	-CC-	- 46	- 34

Cyclohexane

CH		− 39
CCH,HCH		− 77
CC		− 17
CCC	+ 18	−116
−CC−	− 35	− 53

Table 9-8: From PEF400opt to PEF401

	C---C	C---H	H---H
ϵ	0.0596	0.1334	0.3345
r*	3.4827	3.1556	2.7796
A	189775.	130059.	71151.
B	212.70	263.44	308.55
V$\overline{A}$	435.63		266.74
V$\overline{B}$	14.584		17.566
V$\overline{A}$ V$\overline{A}$		116201.	
V$\overline{B}$ V$\overline{B}$		256.18	

Table 9-9: Thermodynamics in the PEF400 series

Substance		PEF400	PEF401opt	PEF304opt	exp. note
Ethane	S	228.9	228.1	227.7	229.5 a
	C	51.5	50.9	49.5	52.8
Propane	S	270.5	268.7	266.2	270.4 a
	C	71.7	70.5	67.5	73.4
Isobutane	S	284.7	293.5	286.9	295.1 a
	C	94.6	92.7	87.7	97.3
Neopentane	S	326.8	307.9	294.8	306.6 a
	C	118.7	116.1	108.8	121.5
n-butane	S	306.1			311.0 a, f
	C	92.3			97.9
n-pentane	S	339.1			350.3 a, f
	C	113.1			120.6
Cyclohexane	S	298.9	297.0	290.2	298.2 b
	C	107.7	105.6	97.1	106.3
Cyclopentane	S		281.3	276.9	292.9 c
n-butane a:g					0.54:0.46
G (g-a)		3.67	4.42	5.27	2.08 d
n-pentane					
H (ag-aa)		3.89			2.43 e

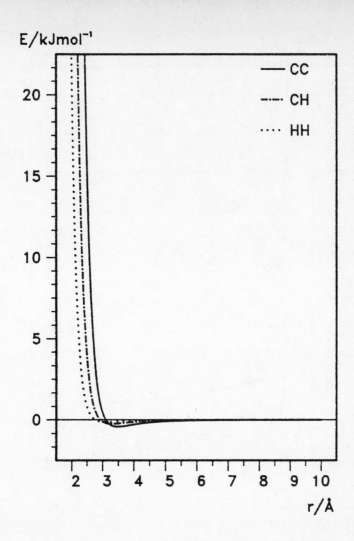

Figure 9-1: Non-bonded part of PEF302opt

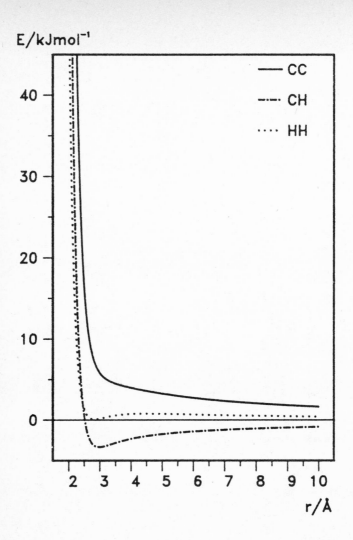

Figure 9-2: Non-bonded part of PEF40lopt

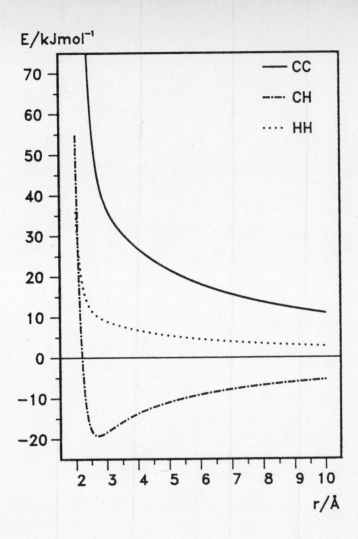

Figure 9-3: Non-bonded part of PEF304opt

9.3 Comparisons

As stated above, there are only tiny differences from one optimised parameter set in the PEF300 series to another. The largest occur for K_ϕ and the non-bonded parameters.

Plots of the various Lennard-Jones functions are very much alike; PEF301opt looks like PEF300[215]; PEF302opt, which is shown in Figure 9-1, is not very different; PEF303opt, which is not shown, is very similar. In short, they are all traditional van der Waals curves.

This just shows that the actual analytical form of the non-bonded functions is of no consequence, and that the inclusion of the unit cell of ethane into optimisation has only marginal effect. When the functional form has little or no importance, the simplest can be used, which is the (A,B) form with one-atom parameters. This conclusion is encouraging for later optimisation of potential energy functions which include heteroatoms.

It is more interesting to compare qualitatively different functions, such as the PEF300 and PEF400 series.

PEF400opt is just like PEF400[171,215,Fig.6]; PEF401opt has the same form but is quantitatively different, see Figure 9-2. Note again the untraditional form, like that of PEF400: The C---C interaction is monotonously repulsive; C---H is attractive far out; H---H has a well-defined minimum and a very broad maximum.

In Figures 9-2 and 9-3 atomic charges were put in as the program calculates them for cyclohexane from the charge parameters.

With few exceptions, PEF401opt gives a better fit than PEF302opt etc. Therefore the test with PEF304 was made: adding charges to PEF303opt gives it the same analytical form as PEF401. The resulting non-bonded functions look a bit stranger, see Figure 9-3; rather like interactions in molecules containing oxygen[171,215,Fig.7]. This indicates what in fact I had expected, that addition of a sizeable term to a potential energy function while optimisation progresses does not ease or even further the work. It must be carefully modelled from the outset.

PEF304 and PEF401 do not converge towards each other which under-
lines what was said above about THE optimum. The output depends on
the input, and the optimisation is not completely objective. Probably no optimisation of this kind is.

Further comparisons between the parameter sets optimised here,
including also PEF304, are given in chapter 10, and a discussion of
their relative merits in section 10.3. Optimisation of PEF303 and
PEF304 is described shortly in section 11.6.2.

The form of the empirical non-bonded functions of the PEF400
series has, qualitatively at least, a theoretical backing[170,256]. Or
vice versa, if the publication dates are considered.

10   Applications: Alkanes and cycloalkanes

Anybody can make a force field for alkanes

Jack D. Dunitz

This chapter will review some of our results obtained on the simpler compounds. I shall describe potential energy functions with parameter sets fitted both by trial and error and by optimisation, and I shall observe chronological order. The older parts include also work on ethers and alcohols.

10.1 Non-optimised potential energy functions

Our consistent force field work developed in a very pragmatic way during the years 1969 - 1979: By simple "hand fitting" we developed a parameter set for coordination complexes, see chapter 4, and one for saccharides, see chapter 5. Both gave good results, and both needed, of course, improvements. We felt an immediate need for a better saccharide parameter set for two reasons: the angle at the oxygen atom in the glycosidic linkage in maltose and cellobiose was too small[168,169], and no account was taken of hydrogen bonding. Therefore we developed the parameter set called PEF400, still by trial and error fitting, but in a more rational way, beginning with simple alkanes and cyclohexane and proceeding to ethers and small alcohols.

The development of this potential energy function has been described in much detail[171], with supplementary notes added later[215]. A further test was performed with a calculation[217] on tris-(tert-butyl)-methane, TTBM, in which the power of the somewhat strange non-bonded parameters was demonstrated. By and large, PEF400, with 16 parameters, gave as good fit as other, more complicated, potential energy functions, with up to 29 parameters.

PEF400 has performed rather well for the purpose for which it was developed, saccharides[172,217], but modelling of hydrogen bonding is not satisfactory in all detail, and for other classes of compounds, some structural details are not accounted for in a satisfactory way.

A crown ether, 18-crown-6, was studied[164] with our earliest potential energy function PEF3[132] applicable to these compounds; the conclusion being that the "internal" part of the potential energy function is too stiff and the non-bonded part not adequate. I have continued those studies, so far unpublished, with PEF300[167] and PEF400. PEF300 gave a slightly better, PEF400 a slightly worse, representation of the set of torsional angles, but the differences are hardly significant. PEF300 gaves unsignificantly better, PEF400 significantly worse, C---O distances. On the other hand, O---O distances are increasingly and significantly more accurate in the sequence PEF3 < PEF300 < PEF400. PEF300 has a more relaxed torsional potential than PEF3, otherwise they are identical; but PEF400 is much more relaxed in several terms and has completely different non-bonded interactions. On this basis it is probably not worthwhile to develop either further, as there is no guidance in this material on how to change parameter values; rather they should be optimised properly, which is about to be done.

Another indication that PEF400 is not satisfactory for all cases comes from a study, also unpublished, of cyclodecane; this will be dealt with below, in connection with an assessment of the newly optimised parameter sets.

10.2 Optimised potential energy functions

The parameter sets PEF301, PEF302, PEF303, PEF304 and PEF401 are documented in chapter 9. They contain parameters for only C and H, and only in saturated compounds.

In addition to the data in Tables 9-2, 9-3, 9-4, 9-6, 9-7 and 9-9, some supplementary data should be mentioned.

10.2.1 Cyclopentane

The calculated rotational constants of cyclopentane show that the molecule is an oblate symmetrical top to within 90 ppm in both PEF302 and PEF303, to 183 ppm in PEF304, and to 3 per mille in PEF401; and the frequency of pseudorotation, which ideally is zero, was calculated to 4, 6, 8 and 22 cm^{-1}, the lowest values ever found when pseudorotation is not considered a special type of motion. The proper torsion was calculated 69 and 40 cm^{-1} too low in PEF302 and PEF303, the two bendings on average 23 and 20 cm^{-1} too high. Cyclopentane gives the largest error in all the thermodynamic calculations, S = 315.4 in PEF303; this is precisely because the pseudorotational frequency is near zero, yet is not excluded from the Einstein sums in ordinary operation of the program. When the usual six external degrees of freedom plus the pseudorotation are excluded, the calculated entropy is closer to the experimental value, see Tables 9-4 and 9-9.

10.2.2 Rotational barriers

Barriers to internal rotation may also be calculated with the consistent force field program. For ethane and n-butane, the procedure is very simple. A geometry is built by the program from standard bond lengths and valence angles[190], with torsional angles specified in the input. Such a geometry has the correct torsions of either the transitional or the minimum conformation, but wrong bond lengths and angles. It is then subjected to steepest descent minimisation, whereby only those atoms which experience a high energy gradient are moved appreciably. The result is, in each case, a conformation with a low energy gradient, close in conformational space to either a minimum or a saddle point, and relaxed in all degrees of

Table 10-1: Barriers to internal rotation

Substance Geometry	E	E(tors)	E(n-b)	CC	CCC	ϕ

PEF302

Ethane

0	12.43	3.05	7.24	1.539		0.001
60	4.31	0.00	3.67	1.529		60.

methyl
barrier	8.12	3.05	3.57			

Butane

60, 0, 60	32.55	3.05	15.93	1.537, 1.558	116.	0.101
60, 60, 60	15.02	0.227	9.24	1.535, 1.541	113.	64.

g - g´
barrier	17.53	2.82	6.69			

PEF303

Ethane

0	14.76	5.36	7.27	1.539		0.001
60	4.36	0.00	3.72	1.529		60.

methyl
barrier	10.40	5.36	3.55			

Butane

| 60, 0, 60 | 35.77 | 5.37 | 16.74 | 1.537, 1.558 | 116. | 0.146 |
| 60, 60, 60 | 16.11 | 0.35 | 10.05 | 1.535, 1.541 | 113. | 64. |

g - g´
| barrier | 19.66 | 5.02 | 6.69 |

| 0,180, 0 | 36.04 | 10.72 | 18.37 | 1.545, 1.540 | 112. | 0., 180. |
| 60,180, 60 | 12.25 | 0.00 | 9.41 | 1.533, 1.537 | 111. | 60., 180. |

methyl
| barrier | 11.90 | 5.36 | 4.48 |

PEF401

Ethane

| 0 | 10.17 | 0.09 | 4.75 | 1.548 | | 0. |
| 60 | 0.71 | 0.00 | -0.86 | 1.531 | | 60. |

methyl
| barrier | 9.46 | 0.09 | 5.61 |

Butane

| 60, 0, 60 | 3.78 | 0.09 | -10.87 | 1.538, 1.559 | 117. | 0.029 |
| 60, 60, 60 | -13.53 | 0.01 | -20.29 | 1.535, 1.539 | 114. | 62. |

g - g´
| barrier | 17.31 | 0.08 | 9.42 |

| 0,180, 0 | 3.57 | 0.19 | -8.56 | 1.551, 1.536 | 114. | 0., 180. |
| 60,180, 60 | -17.37 | 0.00 | -21.46 | 1.532, 1.534 | 112. | 60., 180. |

methyl
| barrier | 10.47 | 0.10 | 6.45 |

| 60,120, 60 | -6.58 | 0.09 | -14.35 | 1.533, 1.552 | 113. | 59., 118. |

| g - a | | | |
| barrier | 6.95 | 0.08 | 5.94 |

| a - g | | | |
| barrier | 10.79 | 0.09 | 7.11 |

----

freedom, including torsions. The degree of relaxation is measured by the norm of the gradient which was $0.003 - 0.005$ kJmol^{-1} A^{-1} for ethane and $0.1 - 2.6$ for butane; in minimisation proper we seldomly accept values above 10^{-6}.

Other indicators of the closeness to saddle point or minimum are the torsional angles, which are almost at their minimum values. Further, the Hessian matrix shows the usual six zero eigenvalues corresponding to the external degrees of freedom, and one, one and two negative values for ethane (0), butane (60,0,60) and butane (0,180,0), giving imaginary frequencies of $180 - 250$ cm^{-1}. The results are shown in Table 10-1.

A revised value for the methyl barrier in ethane was derived from infrared measurements by Weiss and Leroi[274]; their value of 12.25 (0.10) kJmol^{-1} is to be compared with the calculated values 8.12, 10.40 and 9.46 for PEF302, PEF303 and PEF401.

The methyl barrier in butane is higher, 11.90 kJmol^{-1} in PEF303 and 10.47 in PEF401; the experimental value is 13.43 (0.04), as derived by Durig and Compton[67] from vibrational spectra.

The central g-g´ barrier is 17.53, 19.66 and 17.33 kJmol^{-1} in PEF302, PEF303 and PEF401; the energy barrier derived[120] from the heat of formation is 22.2 kJmol^{-1}; from vibrational spectra it is[67] 8.91 kJmol^{-1}.

The two barriers to conformer interchange in butane, g $\rightarrow$ a and a $\rightarrow$ g, were calculated with PEF401 to 6.95 and 10.79 kJmol^{-1}. Estimates from heats of formation[120] give 15.56 kJmol^{-1} for the a $\rightarrow$ g barrier.

As Table 10-1 shows, eclipsing has a marked effect on those parts of the conformation which are usually kept fixed in estimates of barriers. In the g-g´ saddle point, for example, the central CC bond is elongated to 1.558 Å, and the CCC angle is opened to 116°. Similar deformations were found in an ab initio study[32]. The distribution of energy on all internal degrees of freedom is also illustrated by the energy contributions E(torsional) and E(non-bonded) which together account for 0.82 - 0.86 of the methyl barriers and 0.54 - 0.60 of the central barrier, the rest being taken up by bond and angle deformation terms. These observations are in accordance with those of other authors; for a review of experimental and theoretical work on n-butane see Peterson and Csizmadia[204].

10.2.3 Rotational constants

Rotational constants provide a fast but nonspecific test of the goodness of a parameter set. The un-optimised PEF400 is rather good in this respect[171]: average deviations from measured values are 4 to 7 per mille for small alkanes and alkanols. The optimised PEF302 and PEF303 show deviations for alkanes from 3 to 22 per mille, with an average of 10 and 9. In PEF401, the range is 0 to 19, average 5, and in PEF304 it is 3 to 22, average 12.

10.2.4 Hexamethylethane

For a further test of the predictive power of PEF303 and PEF401 with a strained case, hexamethylethane was chosen. The r_g structure is reported by Bartell and Boates[16]: CC 1.542 (peripheral), 1.582 (central), CCC 111.0. The calculated results are, from PEF303: 1.553, 1.594, 111.7, and from PEF401: 1.551, 1.584, 112.7.

Table 10-2: Cyclohexane twist-boat - chair energy difference[a]

	PEF303	PEF304	PEF401
E	22.70	24.75	19.29
E(b)	1.55	1.69	2.59
E(θ)	1.96	1.99	3.40
E(ϕ)	10.06	10.00	0.18
E(Coul)	-	1.77	0.21
E(L-J)	9.14	9.31	12.92
E(n-b)	9.14	11.08	13.12

a) in kJmol^{-1}.

10.2.5 Cyclohexane inversion

The inversion barrier in cyclohexane has received much attention throughout the history of conformational analysis. The enthalpy of activation, as found by NMR[210], is usually 40 - 50 kJmol^{-1} depending on solvent and temperature range.

For the present purpose, testing and comparing parameter sets, there is no need to calculate the conformational space, with all its minima and saddle points; that is available in the literature[206]. A look at two conformers will serve the purpose. The twist-boat is a true minimum energy conformer, and it has an energy somewhat below that of the transition conformation. Table 10-2 shows the difference in total and component energies between twist-boat and chair, as found with PEF303, PEF304 and PEF401.

The results are remarkably alike, the only significant difference being the near disappearance of $E(\phi)$ in PEF401. This is one example of the impossibility of ascribing conformational differences to any one term of a potential energy function; it all depends on the underlying model and its parametrisation. The total energy difference of 20 - 25 kJmol^{-1} corresponds to what is found by other authors using different potential energy functions[7,88].

10.2.6 Cyclodecane

Some years ago, when PEF400 was new, it was applied to cyclodecane. Five conformations were selected. Three of them have been found in crystal structures[65]: Nos. 1, 2 and 4 in Table 10-3; No. 3 was included to have one of another high symmetry, and No. 5 because it had been found by calculation[70]. A crown conformation is not stable to any minimisation algorithm and is therefore not a conformer. The outcome of minimisation was discouraging, and further work postponed till a parameter set had been optimised.

Results of the use of three of the newly optimised sets are shown in Tables 10-3 and 10-4. Inspection of the valence and torsional angles shows that PEF303 and PEF304 give conformer No. 1 the conformation selected by Dunitz[64] for an average C_{2h} ring. The fit of PEF400 and PEF401 is not nearly so good.

The two conformers of C_2 symmetry, Nos. 2 and 4 of Table 10-3, have conformations like those found in the solid state[63].

The relative energies are surprisingly similar in all parameter sets. What is also surprising is the very low energy of the TCCC conformer, lower than that found by Engler, Andose and Schleyer[70]. Table 10-5 shows details of the conformation. In this case, PEF401 gives the same result as the earlier calculation.

Table 10-3: Relative energy[a] of cyclodecane conformers[b]

Conformer				PEF400	PEF303	PEF304	PEF401	ΔG[c]
1	BCB	2323[d]	C_{2h}	29.36	27.49	38.28	32.47	28.00
2	TBC	2233	C_2	25.23	25.83	35.10	28.17	23.81
3	TCBC	122122	D_2	15.54	14.65	18.12	17.29	22.38
4	TBCC	1333	C_2	7.57	2.92	6.34	8.95	6.50
5	TCCC	1414	C_{2h}	0.00	0.00	0.00	0.00	0.00

a) in kJmol^{-1}.

b) Cyclodecane models were built, initial conformations constructed, and minimisations carried out by Birgit Rasmussen.

c) in PEF401.

d) Dale's nomenclature[56].

It seems that a reasonably good parameter set will give a good fit of the conformations of cyclodecane. As to relative stabilities, only conformational energies have been calculated, and they depend very much on the parametrisation; see also the discussion by Burkert and Allinger[45,p.106]. Inclusion of the full statistical mechanical treatment[100,216] does not change the sequence, see Table 10-3, rightmost column.

Table 10-4: The BCB conformer of cyclodecane

	PEF400	PEF303	PEF304	PEF401	Ref.64
θ_1 θ_4 θ_7 θ_9	122	118	118	122	118
θ_2 θ_5 θ_6 θ_{10}	117	115	115	117	114
θ_3 θ_8	120	118	118	120	118
$-\phi_1$ ϕ_4 $\phi_6-\phi_9$	54	54	54	54	55
$-\phi_2$ ϕ_3 $\phi_7-\phi_8$	64	67	67	63	66
$-\phi_5$ ϕ_{10}	144	151	150	144	152

Table 10-5: The TCCC conformer of cyclodecane

	PEF400	PEF303	PEF304	PEF401	Ref.70
θ_1 θ_4 θ_7 θ_9	115	113	112	115	
θ_2 θ_5 θ_6 θ_{10}	118	115	115	118	
θ_3 θ_8	118	114	114	118	
$\phi_1-\phi_4-\phi_6$ ϕ_9	84	87	87	83	83
$-\phi_2$ ϕ_3 $\phi_7-\phi_8$	142	144	145	142	144
$\phi_5-\phi_{10}$	69	72	73	69	68

10.3 Reliability

A prospective user of a potential energy function is first of all interested in its reliability. I hope to have shown by my examples in sections 9.1 and 9.2 and in this chapter that a parameter set may be good or just good enough for some purposes but insufficient for others. This applies also to optimised sets and does not mean that a set is not consistent. Rather, the model behind the potential energy function is too primitive, or the parametrisation behind the parameter set is not well chosen for the purpose.

Reliability has two aspects: reproducibility and predictability. A short summary of what we had obtained with un-optimised potential energy functions was published recently[219]. On basis of the results given in chapter 9 and here it can be said that optimisation has in general improved the postdiction of data used in optimisation, namely conformations and vibrational spectra; that the postdictive capability of the PEF300 series for data not used in optimisation (technically this ia a predictive capability) has improved; that the opposite is the case for the PEF400 series, though to a small extent, and mostly for highly strained systems.

The conclusion is therefore that PEF302, PEF303 and PEF304 are more reliable for prediction than PEF401, and that good conformations and thermodynamic properties may be expected from the former, that conformations from PEF304 are best, at the expense of thermodynamic functions.

11   Extension to crystals

by Kjeld Rasmussen and Lars-Olof Pietilä

The glory of young men is their strength,
the dignity of old men their grey hair.

Proverbs XX, 29

Optimisation also on crystal structures is desirable, despite the considerably increased complexity of the programs, because crystal packing is determined exclusively by non-bonded interactions; conversely, therefore, crystal packing is eminently suitable for optimisation of non-bonded energy function parameters.

The usual way to sum non-bonded interactions over a crystal lattice is very simple and very time-consuming: sum within a unit cell; between this and its nearest-neighbours; include nextnearest-neighbours etc.; using a summation limit of anything between 6 and 50 A[266].

A more rational way is to use convergence acceleration, known to solid-state physicists for a lifetime, but only reluctantly adopted by chemists. Convergence acceleratation was introduced into the treatment of molecular crystals by Williams[280]. The choice of convergence constant for the attractive potential was further studied by Mirsky[174]. To our knowledge, nobody has yet used Williams´ methods in the full CFF context, and a paper by us [207] is probably the first attempt to do so.

For an introduction to and review of convergence acceleration, we refer the reader to Williams´ recent paper[284]. We shall here keep to Williams´ nomenclature where possible and practical, and shall give the results of all derivations, and all formulae necessary for implementation. They differ in places from those of Williams. We also describe the changes and additions made to our consistent force field program, which means that the bulk of this chapter, just as chapter 7, functions as a sequel to to the previous book [190].

Optimisation will be done in the CFF context, and we shall therefore be at variance with Williams´ formulation.

We have selected a few cases of molecular and ionic crystals as
tests for our new program. They include crystals of Ar, C_2H_6, KCl
and NaCl. We shall show the significance of using convergence acceleration also on short-range repulsion terms, and the interplay of
inter- and intramolecular forces in crystals of flexible molecules,
an aspect which is usually given insufficient attention.

Finally, we shall recount our first experiences with optimisation
on unit cell dimensions and lattice energy.

11.1   Methods

As we use non-rigid molecules, we have to calculate both the
intramolecular and the intermolecular contributions to the energy
and energy derivatives. As variables we have the six lattice con-
stants (a, b, c, α, β and γ), and the 3N position coordinates of the
N atoms in the asymmetric unit.

11.1.1   Coordinate systems

The position coordinates are given in cartesian coordinates
rather than in fractional cell coordinates, making it much easier to
calculate the intramolecular energy. It is also advantageous that
the molecular geometry remains unchanged when the lattice constants
are varied.

We need a reference cartesian coordinate system. Following Wil-
liams[280,284], we take $\underline{e}_y$ and $\underline{e}_z$ coincident with $\underline{b}$ and $\underline{c}^*$
respectively, where $\underline{c}^* = \underline{a} \times \underline{b} / \underline{a} \cdot (\underline{b} \times \underline{c})$ is the reciprocal lattice
vector. $\underline{e}_x$ is then in the $\underline{a}, \underline{b}$ plane making the coordinate system

righthanded. The cartesian coordinates of the asymmetric unit are given in a shifted coordinate system, the origin being at the point $\underline{y}$, where $\underline{y}$ is given in fractional cell coordinates. This allows us to choose the orgin in the most convenient way, for example in a center of symmetry of the asymmetric unit. The axes of both systems are parallel. Here we differ from Williams. The cartesian coordinates of the atom j in the reference cartesian coordinate system are

$$\underline{X}_j = \underline{x}_j + D\underline{y}$$

where $\underline{x}_j$ are the cartesian coordinates in the shifted coordinate system, and the matrix D, which transforms the fractional cell coordinates to cartesian coordinates, is

$$D = \begin{bmatrix} a \sin \gamma & 0 & c(\cos \beta - \cos \alpha \cos \gamma)/\sin \gamma \\ a \cos \gamma & b & c \cos \alpha \\ 0 & 0 & c \delta/ \sin \gamma \end{bmatrix}$$

where $\delta = (1-\cos^2\alpha -\cos^2\beta-\cos^2\gamma+2\cos\alpha \cos\beta\cos\gamma)^{1/2}$.

The other asymmetric units are generated from the original asymmetric unit by symmetry operations. Let $\hat{\underline{X}}_j$ be the cartesian coordinates of the atom j in the generated asymmetric unit. Then

$$\hat{\underline{X}}_j = R\left\{\hat{\underline{x}}_j + D(\underline{y}-\underline{g}_R)\right\} + D(\underline{g}_R+\underline{T}_R),$$

where R is the (symmetry) transformation matrix, $\underline{T}_R$ the translation vector, and $\underline{g}_R$ specifies the position of the symmetry element. Note that $\underline{g}_R$ and $\underline{T}_R$ are given in fractional cell coordinates. The vector $\underline{r}_{kj}$ pointing from the atom k in the original asymmetric unit to the atom j in the generated asymmetric unit is

$$\underline{r}_{kj} = R D(\underline{y}-\underline{g}_R)+ D(\underline{g}_R+\underline{T}_R-\underline{y})+ R\hat{\underline{x}}_j-\underline{x}_k.$$

11.1.2 Derivatives

The dependence of $\underline{r}_{kj}$ on the lattice constants p_ν ($\nu = 1, \ldots, 6$) comes in the first two terms through $\mathbb{D}$; dependence on the position coordinates is found in the last two terms. The derivatives of $\underline{r}_{kj}$ with respect to lattice constants are

$$\frac{\partial \underline{r}_{kj}}{\partial p_\nu} = R\left(\frac{\partial \mathbb{D}}{\partial p_\nu}\right)(\underline{y}-\underline{q}_R)+\left(\frac{\partial \mathbb{D}}{\partial p_\nu}\right)(\underline{q}_R+\underline{T}_R-\underline{y})$$

$$\frac{\partial^2 \underline{r}_{kj}}{\partial p_\nu \partial p_\mu} = R\left(\frac{\partial^2 \mathbb{D}}{\partial p_\nu \partial p_\mu}\right)(\underline{y}-\underline{q}_R)+\left(\frac{\partial^2 \mathbb{D}}{\partial p_\nu \partial p_\mu}\right)(\underline{q}_R+\underline{T}_R-\underline{y})$$

Note that equations (5) correspond to Williams´ appendix[284].

The derivatives of $\underline{r}_{kj}$ with respect to the position coordinates are

$$\frac{\partial \underline{r}_{kj}}{\partial \hat{\underline{x}}_j} = \mathbb{R} \qquad \text{and} \qquad \frac{\partial \underline{r}_{kj}}{\partial \underline{x}_k} = -\mathbf{1} \qquad \text{when} \qquad k \neq j$$

$$\frac{\partial \underline{r}_{jj}}{\partial \underline{x}_j} = \mathbb{R} - \mathbf{1}.$$

The derivatives of the distance r_{kj} are then

$$\frac{\partial r_{kj}}{\partial p_\nu} = \frac{1}{r_{kj}}\sum_{\gamma=1}^{3} r_{kj,\gamma}\frac{\partial r_{kj,\gamma}}{\partial p_\nu}$$

$$\frac{\partial r_{kj}}{\partial x_{j\alpha}} = \frac{1}{r_{kj}}\sum_{\gamma=1}^{3} r_{kj,\gamma}\, R_{\gamma\alpha} \qquad k \neq j$$

$$\frac{\partial r_{kj}}{\partial x_{k\alpha}} = - \frac{r_{kj,\alpha}}{r_{kj}} \qquad k \neq j$$

$$\frac{\partial r_{jj}}{\partial x_{j\alpha}} = \frac{1}{r_{jj}} \sum_{\gamma=1}^{3} r_{jj,\gamma} \left(\mathbb{R}^{-1} \right)_{\gamma\alpha}$$

$$\frac{\partial^2 r_{kj}}{\partial p_\nu \partial p_\mu} = \frac{1}{r_{kj}} \sum_{\gamma=1}^{3} \left\{ r_{kj,\gamma} \left(\frac{\partial^2 r_{kj,\gamma}}{\partial p_\nu \partial p_\mu} \right) + \left(\frac{\partial r_{kj,\gamma}}{\partial p_\nu} \right) \left(\frac{\partial r_{kj,\gamma}}{\partial p_\mu} \right) \right\}$$

$$- \frac{1}{r_{kj}} \left(\frac{\partial r_{kj}}{\partial p_\nu} \right) \left(\frac{\partial r_{kj}}{\partial p_\mu} \right)$$

$$\frac{\partial^2 r_{kj}}{\partial \hat{x}_{j\alpha} \partial x_{k\beta}} = - \frac{1}{r_{kj}} \left\{ R_{\beta\alpha} + \left(\frac{\partial r_{kj}}{\partial x_{k\beta}} \right) \left(\frac{\partial r_{kj}}{\partial \hat{x}_{j\alpha}} \right) \right\} \qquad k \neq j$$

$$\frac{\partial^2 r_{kj}}{\partial \hat{x}_{j\alpha} \partial \hat{x}_{j\beta}} = \frac{1}{r_{kj}} \left\{ \delta_{\alpha\beta} - \left(\frac{\partial r_{kj}}{\partial \hat{x}_{j\alpha}} \right) \left(\frac{\partial r_{kj}}{\partial \hat{x}_{j\beta}} \right) \right\} \qquad k \neq j$$

$$\frac{\partial^2 r_{kj}}{\partial x_{k\alpha} \partial x_{k\beta}} = \frac{1}{r_{kj}} \left\{ \delta_{\alpha\beta} - \left(\frac{\partial r_{kj}}{\partial x_{k\alpha}} \right) \left(\frac{\partial r_{kj}}{\partial x_{k\beta}} \right) \right\} \qquad k \neq j$$

$$\frac{\partial^2 r_{jj}}{\partial x_{j\alpha} \partial x_{j\beta}} = \frac{1}{r_{jj}} \left\{ \sum_{\gamma=1}^{3} \left(\mathbb{R}^{-1} \right)_{\gamma\alpha} \left(\mathbb{R}^{-1} \right)_{\gamma\beta} - \left(\frac{\partial r_{jj}}{\partial x_{j\alpha}} \right) \left(\frac{\partial r_{jj}}{\partial x_{j\beta}} \right) \right\}$$

$$\frac{\partial^2 r_{kj}}{\partial p_\nu \partial \hat{x}_{j\alpha}} = \frac{1}{r_{kj}} \left\{ \sum_{\gamma=1}^{3} \left(\frac{\partial r_{kj,\gamma}}{\partial p_\nu} \right) R_{\gamma\alpha} - \left(\frac{\partial r_{kj}}{\partial \hat{x}_{j\alpha}} \right) \left(\frac{\partial r_{kj}}{\partial p_\nu} \right) \right\} \qquad k \neq j$$

$$\frac{\partial^2 r_{kj}}{\partial p_\beta \partial x_{k\alpha}} = -\frac{1}{r_{kj}} \left\{ \left(\frac{\partial r_{kj,\alpha}}{\partial p_\beta} \right) + \left(\frac{\partial r_{kj}}{\partial x_{k\alpha}} \right) \left(\frac{\partial r_{kj}}{\partial p_\beta} \right) \right\} \qquad k \neq j$$

$$\frac{\partial^2 r_{jj}}{\partial p_\beta \partial x_{j\alpha}} = \frac{1}{r_{jj}} \left\{ \sum_{\gamma=1}^{3} \left(\frac{\partial r_{jj,\gamma}}{\partial p_\beta} \right) (\mathbb{R}-\mathbb{1})_{\gamma\alpha} - \left(\frac{\partial r_{jj}}{\partial p_v} \right) \left(\frac{\partial r_{jj}}{\partial x_{j\alpha}} \right) \right\}$$

In these equations, α, β = 1, 2, 3.

11.1.3 Lattice summation

The intermolecular potential energy in molecular crystals is usually approximated by intermolecular atom-atom potentials, and the
molecules are usually treated as rigid bodies. When the molecules in
molecular crystals are treated as non-rigid, the most logical
approach is to add the intermolecular contributions using the same
potential functions and parameters as for free molecules.

The lattice summation is very time consuming. For that reason the
summation limit has usually been chosen to 8Å or even less, resulting in a rather large termination error, often 100 per mille.
Kitaigorodsky[136] has shown that a summation limit as large as 15Å is
needed to achieve 10 per mille accuracy. This approach needs too
much computer time to be practical. For that reason methods to
accelerate the convergence of lattice sums have been developed,
especially by Williams[280,284], based on work by Nijboer and de
Wette[185]. In this method the lattice sum of form

$$S_n = \frac{1}{2} \sum_{i \neq j}^{\text{lattice unit cell}} \sum_{j} q_i q_j r_{ij}^{-n}$$

is divided into two terms by using a properly chosen convergence
function $\phi(r)$. We give here only the results; the derivation of
equation (14) is given by Williams[280,284].

$$S_n = \frac{1}{2\Gamma(n/2)} \left\{ \sum_i \sum_j q_i q_j r_{ij}^{-n} \Gamma(n/2, a^2) \right.$$

$$+ \frac{2\pi^{n/2} K^{n-3} Z}{(n-3) V_{cell}} \left(\sum_j q_j \right)^2 - \frac{2K^n \pi^{n/2}}{n} \sum q_j^2$$

$$+ \frac{\pi^{n-3/2}}{V_{cell} Z} \sum_{\underline{h} \neq \underline{0}} |F(\underline{h}_\lambda)|^2 h_\lambda^{n-3} \Gamma((-n+3)/2, b^2) \Big\}$$

intramol.contacts
in asym.unit

$$- \sum_{i>j} \sum_j q_i q_j r_{ij}^{-n} (1-\phi(r_{ij}))$$

where $a^2 = \pi K^2 r_{ij}^2$, $b^2 = \pi h_\lambda^2 / K^2$,

$\phi(r) = \Gamma(n/2, K^2 \pi r^2)/\Gamma(n/2)$.

$\Gamma(m, \xi) = \int_\xi^\infty dt\, t^{m-1} e^{-t}$ is the incomplete gamma function, K is the convergence constant, $\underline{h}$ is the reciprocal space vector, $F(\underline{h})$ is the structure factor and Z is the number of asymmetric units per unit cell. If K is chosen properly, the sum in reciprocal space can be neglected, while the sum in direct space still converges rapidly[280].

Equation (14) is different from two of the three versions given by Williams in three different papers[280,284,281], which contain misprints. The errors pertain to the use of Z. In the first of these papers, Williams multiplies $\sum q_j^2$ by Z; it should be $(\sum q_j)^2$. In the second, there are analogous errors :

$(\sum_\alpha A_{\alpha\,\alpha}^{1/2})^2$

should be multiplied by Z, not

$(\sum_\alpha A_{\alpha\,\alpha})$ and $(\sum_\alpha q_\alpha^2)$.

In the third, in equation (2.61),

$$(\sum_{\text{asym}} A_{jj}^{1/2})^2$$

should be multiplied, not divided, by Z.

The text following equation (2.60) is correct. For the cases n = 1, 6, 9 and 12 the sums S_n are given by

$$S_1 = \frac{1}{2} \sum_i \sum_j q_i q_j r_{ij}^{-1} \text{ERFC}(a)$$

$$- K \sum_j^{\text{asym unit}} q_j^2 - \sum_{i>j} \sum_j q_i q_j r_{ij}^{-1} \text{ERF}(a)$$

$$S_6 = \frac{1}{2} \sum_i \sum_j q_i q_j r_{ij}^{-6} (1+a^2+a^4/2) e^{-a^2}$$

$$+ \frac{\pi^3 K^3 Z}{6 V_{\text{cell}}} (\sum_j q_j)^2 - \frac{\pi^3 K^6}{12} \sum_j q_j^2$$

$$- \sum_{i>j}^{\text{asym unit}} \sum_j q_i q_j r_{ij}^{-6} \left\{ 1-(1+a^2+a^4/2) e^{-a^2} \right\}$$

$$S_9 = \frac{1}{2} \sum_i \sum_j q_i q_j r_{ij}^{-9} \left\{ \frac{1}{\sqrt{\pi}}(\frac{16}{105}a^7+\frac{8}{15}a^5+\frac{4}{3}a^3+2a) e^{-a^2} \right.$$

$$\left. + \text{ERFC}(a) \right\} + \frac{8\pi^4 K^6 Z}{315 V_{\text{cell}}} (\sum_j q_j)^2 - \frac{16\pi^4 K^9}{945} \sum_j q_j^2$$

asym unit
$$-\sum_{i>j}\sum_{j} q_i q_j r_{ij}^{-9} \left\{ ERF(a) - \frac{1}{\sqrt{\pi}}(\frac{16}{105}a^7 + \frac{8}{15}a^5 + \frac{4}{3}a^3 + 2a)e^{-a^2} \right\}$$

$$S_{12} = \frac{1}{240}\sum_{i}\sum_{j} q_i q_j r_{ij}^{-12}(a^{10} + 5a^8 + 20a^6 + 60a^4$$

$$+ 120a^2 + 120)e^{-a^2} + \frac{\pi^6 K^9 Z}{1080V_{cell}}(\sum_{j} q_j)^2$$

asym unit
$$-\frac{\pi^6 K^{12}}{1440}\sum_{j} q_j^2 - \sum_{i>j}\sum_{j} q_i q_j r_{ij}^{-12} \left\{ 1 - (a^{10} + 5a^8 + 20a^6 \right.$$

$$\left. + 60a^4 + 120a^2 + 120)e^{-a^2}/120 \right\}$$

In these equations j runs over the asymmetric unit and i runs over the atoms in the surrounding molecules, except where otherwise stated. When both i and j run over the asymmetric unit, only the intramolecular interactions are counted. ERF(x) and ERFC(x) are the error function and the complementary error function, respectively.

11.2 Implementation

As the consistent force field program has a modular structure, changes in or additions to energy processing subprograms is easy. Implementing crystal calculations involve more than just changing addressing, though this alone was an intricate and time-consuming part of the programming.

Referring to the general structure described before[190], changes were made to the treatment of input formula, essentially in BRACK, and to the printout from CONFOR.

In minimisation routines, changes of addressing were made (in STEEPD and STEPSZ). This was necessary because selected lattice constants may have to be kept unchanged when they are used in symmetry operations, such as two angles of a monoclinic crystal.

A section of 13 new routines for energy calculation, coordinate transformation, and addressing, was written; it works parallel to the old sections for bond, angle, non-bonded, etc., interactions, as one more branch under MOLEC. In addition, we use the NAG routines S15ADF and S15AEF for ERFC and ERF.

Convergence acceleration is used on all terms of non-bonded interactions. These terms are expressed as powers of interatomic distances: r^{-12}, r^{-9} (not yet tested in full), r^{-6}, r^{-1}. On terms in r^{-12} and r^{-9} convergence acceleration may be switched off without program changes. Energy minimisation is performed with the same algorithms as in the "standard" CFF (steepest-descent, Davidon-Fletcher-Powell, modified Newton).

Implementation of the changes and additions was made so that the program performs as usual for isolated molecules. For instance, energy minimisation and normal coordinate analysis of gaseous ethane and butane may be done in the same run as energy minimisation of crystalline ethane, with the same energy function parameters.

Input requirements have been kept as small as possible. Where (CH3CH3) in the old program would give all possible interactions in the molecule it still does so in the new; ((CH3CH3)) or ((M)(+1)(X)(-1)) now defines an asymmetric unit of a crystal. Information about the unit cell dimensions, number of asymmetric units per unit cell, symmetry elements, and cartesian coordinates of the asymmetric unit are then also required.

As a spin-off, molecular complexes may be treated, for example the water dimer ((OH2)2) and glycine in a water-shell ((NH3CH2KQ2)(OH2)18).

11.3 Test calculations

11.3.1 Argon

We first tested our program on the Ar crystal, which is fcc, with a = 5.312 A[209]. The potential function used was a Lennard-Jones 12-6 given by Brown[36]; with $(\varepsilon, r^*) = (165.0*10^{-16}$erg, 3.819 A) for a pair of Ar atoms, or (A, B) = (1512, 38.39) for one atom, with units such that energy is in kcalmol^{-1} and distance in A, as required by the CFF program.

Three summation limits were used, 6, 8 and 10 A, each with a range of values of the convergence constant K, different values being used for the repulsive and attractive terms. Our previous paper[207] gives a full discussion of our investigation of various combinations.

If K should be the same for repulsive and attractive terms, preferred values are about 0.23, 0.20 and 0.17 A^{-1} for summation limits of 6, 8 and 10 A. Proper choice of K is much more critical for a small summation limit, as would be expected. For the Ar case, a good compromise would be a summation limit of 8 A and K = 0.20 A^{-1}.

The effect of convergence acceleration on the repulsive term is not large, but significant for a summation limit of 6 A, and negligible for 8 and 10 A. A K as large as 0.30 A^{-1} is unacceptable when summation in reciprocal space is left out.

11.3.2 Ethane

The structures of ethane at low temperature were described in great detail by van Nes and Vos[184]. We use the monoclinic structure of the low-temperature form, and coordinates from their model B1, transformed into cartesians. This means a C-C bond length of 1.506

A, C-H bond lengths ranging from 0.941 A to 1.003 A, and H-C-H angles ranging from 103 to 111°.

Although this geometry looks unreasonable, we have used it because it was calculated without constraints[184], and because it might give us an initial point in the combined crystal-plus-conformation space which would not be very close to minimum, so as to present a real task to the minimiser.

We used two potential energy functions. One is a modification of an earlier one, called PEF400, which has worked well for many types of molecules[172], even the overcrowded tris- (tert-butyl)-methane[217] where non-bonded interactions play an overwhelming role within the molecule, as they do between the molecules in a crystal.

The non-bonded interactions in PEF400 are given as a sum of Lennard-Jones 12-6 and Coulomb interactions, the pure Lennard-Jones curves having rather deep minima. In the fitting of energy function parameters, a dielectric constant of 3 was used.

The Lennard-Jones parameters of PEF400 are given as ε and r^* for pairs of atoms. They were transformed to A and B corresponding to

$$V = \varepsilon(r^*/r)^{12} - 2\varepsilon(r^*/r)^6 = Ar^{-12} - Br^{-6},$$

and in the new PEF403 $\sqrt{A}$ and $\sqrt{B}$ for C---C and H---H interactions were used as single-atom parameters q_1.

For atomic charges we used the values assigned by the program from the parameter values of PEF400[172].

The other function, PEF303, was a modification of the well-tried PEF300. The original exp-6 functions were changed into the 12-6 form with (ε, r^*) parameters, PEF301[218], and (A,B) single atom parameters for PEF303 were found as just described for PEF403.

All parameter values are shown in Table 11-1.

In our paper on crystal energy minimisation[207] we recount at length our choices of convergence constant, summation limit and initial structure. Our conclusion there is that we may safely use a summation limit of 6 A, with a convergence constant of 0.20 A^{-1}.

Table 11-1: Energy function parameters for ethane.[a,b]

Interaction	K_b, K_θ, A, e	b_0, θ_0, B
Modified PEF400: PEF403		
C-C	510.	1.509
C-H	670.	1.093
C-C-H	71.	arcos(-1/3)
H-C-H	75.	arcos(-1/3)
C--	581.3	19.17
H--	236.9	16.11
C.	-0.4125	
H.	+0.1375	
Modified PEF301: PEF303		
C-C	720.	1.52
C-H	720.	1.09
C-C-H	93.5	1.911
H-C-H	74.8	1.911
C--	573.6	18.95
H--	131.7	7.046

a   Units: K_b, kcal mol^{-1} A^{-2};
K_θ, kcal mol^{-1} rad^{-2};
b_0, A; θ_0, rad;
A, sqrt(kcal mol^{-1} A^{12});
B, sqrt(kcal mol^{-1} A^{6});
e, elementary charge

b   Symbols as in ref. 190

Table 11-2: Minimised structures of ethane

	PEF403		PEF303		Experimental
Coordinate	Crystal	Free molecule	Crystal	Free molecule	x-ray, B3
C–C	1.528	1.530	1.527	1.527	(1.532)
C–H(1)	1.091		1.090		
C–H(2)	1.087	1.096	1.089	1.091	(1.096)
C–H(3)	1.089		1.090		
C–C–H(1)	111.7		110.2		109.5
C–C–H(2)	111.2	110.7	110.1	110.0	111.8
C–C–H(3)	111.1		110.1		108.8
H(1)–C–H(2)	107.2		108.8		107.1
H(1)–C–H(3)	108.2	108.2	108.9	108.9	111.4
H(2)–C–H(3)	107.3		108.8		108.3
a	4.035		4.229		4.226
b	5.378		5.472		5.623
c	5.531		5.607		5.845
B	93.85		92.5		90.41
<(C–C,a)	102.7		103.1		102.0
<(C–C,b)	47.3		47.1		45.2
<(C–C,c)	133.3		133.3		133.1
Lattice energy*/kJmol^{-1}	-104.0		-26.0		-22.0

* The experimental lattice energy is calculated from the heat of sublimation[136,p.335] using the equation $-E_{latt} = H_{subl} + 2RT$. The calculated lattice energies are corrected for the difference of the intramolecular contribution in the lattice and strain energy in the free molecule.

Minimisation of the total energy of the ethane crystal is a very difficult task, and therefore it presents a real test of the program. The reason is that the structure is rather loosely packed[184], so that the minimum is very shallow; in common terms: the ethane crystal is very soft. For minimisation we used both potential energy functions. In the minimisations we had no other constraints than what crystal symmetry requires.

It is seen from Table 11-2 that the ethane crystal structure is reproduced to within 60 per mille in unit cell dimensions. Note that the two potential energy functions were developed for free molecules. Both will be acceptable candidates for optimisation, but should not be used in general in their present form.

PEF403 has very hard non-bonded interactions and gives a structure which is too small and has much too large lattice energy. Even this function could not reproduce the large differences in bond angles between crystal and gas phases, up to 4°, which van Nes and Vos calculated for their structure B3.[18] This indicates that the ethane molecule is not so distorted in the crystal as the structure B3 suggests. PEF303, which has rather soft non-bonded interactions, gives no significant distortion, and a lattice energy close to the experimental.

To study the effect of constraints we also minimised the total energy, constraining the angle B to the experimental value. In a second minimisation, all the lattice constants were constrained to their experimental values. PEF303 was used in both of these calculations. The first gave only an insignificant decrease (about 4 per mille) in lattice energy, and yet the lattice constants a, b, and c came out as 4.272, 5.480, and 5.550 A. The orientation of the C-C axis was 103.8, 46.0, and 132.7° (see Table 4). The second gave a decrease of 35 per mille in lattice energy. The orientation of the C-C axis was 101.5, 43.6, and 131.3°. These results also indicate the shallowness of minima.

11.3.3 Potassium chloride and sodium chloride

As a third test case, we wanted an ionic compound, where the Coulomb interaction is dominating. We chose KCl because we wanted to use Ar parameters in the van der Waals terms (the Lennard-Jones functions). NaCl was also used in the convergence tests.

The behaviour of the energy when K is changed is not analogous to the previous two cases, as we here have charges of alternating sign throughout the lattice. Therefore we must compare the calculated Coulomb energy with the energy calculated from the Madelung constant and the unit cell edge (often called the Madelung energy). It should here be stated again that the CFF program usually assumes a dielectric constant of 3; runs with D = 1 were made for the test cases reported here.

Our previous results[207] show that the proper range of K is from about 0.15 to 0.20 A^{-1}, and that if there are strong Coulomb interactions we should use a summation limit of 8 A or more.

To test the minimisation in the predominantly ionic case, we used the argon parameters for the Lennard-Jones potentials in KCl. Convergence constants K_{Coul} = 0.15 A^{-1}, K_{attr} = K_{rep} = 0.20 A^{-1} and summation limit = 8 A were used. In the minimum, a = 6.052 A, E_{Coul} = -802.6 kJmol^{-1}, and E_{VdW} = 33.2 kJmol^{-1}. The cell edge is about 40 per mille shorter than the experimental. It is a bit surprising that the difference is not larger, as the model is very crude, and the nearest neighbour distance in KCl (3.15 A) is much shorter than in the argon crystal (3.76 A).

11.4 Optimisation on crystals

The lattice energy of a crystal is given as a sum of repulsive, attractive, and Coulombic interactions. We have chosen not to optimise charge parameters on lattice observables. There are two main reasons for that.

Firstly, the microscopic dielectric constant is always uncertain, so that the charge parameters would be uncertain. For example, Williams[282] optimised charge separation in the C-H bond along with other potential energy function parameters on lattice variables of several hydrocarbons. The optimum value for the charge separation gave a bond moment of 0.88 D for the C-H bond whereas the commonly used value is about 0.4 D [282].

Secondly, heteroatoms and different bond types would cause many difficulties.

Therefore, when optimising on lattices observables, we have until now chosen to optimise only the parameters for repulsive and attractive interactions. To optimise on lattice energy, we use crystals where the Coulombic contribution is small, when possible. Like Williams[282], we prefer calorimetric data for heats of sublimation to data derived from vapour pressure measurements. Unfortunately the latter are much more common in the literature. A discussion of the calculation of lattice energy from experimental data is given by Shipman et. al.[245].

In addition, in flexible molecules we have to take into account the difference of the strain energy in the crystalline and gas phases. This is usually small, and can be approximated by the difference calculated by the initial potential energy function. The goodness of this approximation is easily checked using the final potential energy function.

11.4.1 Choice of algorithm

There has been some discussion about the residual to be minimised [92,152,250,283,284]. Williams has chosen to minimise the forces at the experimental structure, which he calls the force fit method. This method has some advantage over the method of Hagler and Lifson, which minimises the difference between the observed and estimated structures; Williams calls it the direct parameter fit method.

Both methods give equally good fits to the structure, but the parameters from the former give a much better estimate of lattice vibrations[250].

There are also some drawbacks in the force fit method. In shallow minima, the forces may be small although the distance from the minimum is large. Hagler and Lifson[92] found that the forces and the distances from the actual energy minimum do not necessarily correlate. Another difficulty in the force fit method is that there must be at least one lattice energy to optimise on, as otherwise the potential energy function parameters for non-bonded interactions may go to zero to minimise the forces[284]. In hydrocarbons this is not a problem, but sometimes it is difficult to find a reliable experimental value for the lattice energy.

Also the direct parameter fit method has drawbacks. Lifson and Levitt[152], citing Hagler, state that final parameter values occasionally depend on the choice of initial values which might indicate a local rather than a global minimum of the residual.

Such a possibility must of course always be considered, but in our opinion this shortcoming is due basically to the approximations inherent in the method, see eqn. 8 and Table 2 of the paper by Hagler and Lifson[92].

11.4.2 Derivatives in parameters

For optimisation on lattice observables we need partial derivatives of each type of observable with respect to potential energy function parameters. So far, we have two types of observables: conformation, including lattice constants, and lattice energy.

Optimisation on molecular conformation is done in all details as for free molecules. Technically we treat lattice constants as up to six extra conformational observables; the partial derivatives are gotten exactly as for derivatives in cartesian coordinates, see section 7.3.1.

The lattice energy is made up of lattice sums S_n. In S_n we have terms of the form

$$q_i(u)q_j(v)f(r_{ij}).$$

where $f(r_{ij})$ is a function of the distance between atoms i and j, and u and v are indices for the types of atoms. $f(r_{ij})$ is independent of potential energy function parameters. The contribution of terms of this type to $\partial S_n/\partial q(w)$ is

$$(\delta_{uw}q_j(v) + \delta_{vw}q_i(u))\ f(r_{ij}).$$

When convergence acceleration is used we have also terms which depend on $(\sum q_j(u))^2$ and on $\sum q_j(u)^2$. The contributions of these to $\partial S_n/\partial q(w)$ are

$$2A\ n_w \sum_j q_j(u) \text{ and } 2B\ n_w\ q(w),$$

respectively, where n_w is the number of atoms of type w in the asymmetric unit, B is a constant, and A depends on lattice constants through the unit cell volume. These equations are used in the calculation of lattice energy derivatives with respect to potential energy function parameters.

In the methods of Williams[284] and of Hagler and Lifson[92] it is necessary to be able to write the lattice sums in this way, as they, and therefore the functions $f(r_{ij})$, are calculated only once, namely at the experimental geometry.

In lattice energy optimisation we use the same formulation as the other two groups, but calculate the derivatives after each minimisation, at the calculated geometry, that is once per parameter per cycle of optimisation. This is much more accurate and much more costly.

11.5 Implementation

In our program we use the direct parameter fit method, but in each cycle we minimise the energy, and calculate the deviations from experimental values as for other observables. This method has the drawback that lattice sums must be calculated repeatedly, which is time consuming, but by using convergence acceleration it is possible in a reasonable computer time. Later, when we intend to optimise on lattice vibrations, we shall have to do it in this way, since the derivatives of lattice vibrations with respect to potential energy function parameters must be calculated numerically[250]. It should be remarked that these calculations are done only when we develop new non-bonded potential energy function parameters.

11.6 Test calculations

Since optimisation on lattice variables is time consuming, we have chosen as test cases Ar, KCl and ethane, which are cheap in computer time and yet go through all the new parts of the program.

Table 11-3: Optimisation on Ar and KCl

	Initial	Test 1	Test 2	Exp.
Ar				
a (Å)	5.245	5.359[a]	5.318[a]	5.312
$-E_{latt}/kJmol^{-1}$	8.523	6.568	7.21[a]	7.74
KCl				
a (Å)	6.052[a]	6.080[a]	6.070	6.293
$-E_{latt}/kJmol^{-1}$	769.4	761.2	764.0	
$-E_{Madelung}$	802.6	798.9	800.2	771.63[b]
Parameters				
A^c	1512.	1513.5	1512.4	
B^c	38.39	35.99	36.82	

[a] Included in optimisation.
[b] Calculated from the Madelung constant.
[c] Units of parameters: A: $(kcalmol^{-1} Å^{12})^{1/2}$.
    B: $(kcalmol^{-1} Å^{6})^{1/2}$.

11.6.1 Argon and potassium chloride

In the first test we optimised Lennard-Jones parameters for Ar, K^+ and Cl^- on lattice constants for Ar and KCl crystals. We used, as described before, single-atom rather than atom pair potential energy functions, and we constrained both the repulsive and the attractive terms to be equal for all three atoms.

The experimental lattice constants for Ar and KCl are 5.312 Å[209] and 6.293 Å[275] respectively. Initial values for the parameters were $A = 1512$ (kcalmol^{-1} Å^{12})$^{1/2}$ and $B = 38.39$ (kcalmol^{-1} Å^6)$^{1/2}$. In all optimisation tests we used the following summation limits and convergence constants: 6 Å, $K_{attr} = 0.23$ Å^{-1} (Ar), and 8 Å, $K_{attr} = 0.20$ Å^{-1} (KCl). Convergence acceleration on the repulsive potential was not used in these tests. The dielectric constant was set to 1, and K and Cl were given the charges +1 and -1. The optimum was reached after 6 iterations; the results are given in Table 11-3.

In the second test we optimised also on the lattice energy of Argon, which is -7.740 kJ mol^{-1} (-1.850 kcal mol^{-1})[36]. It was given a weight 0.3 times the weight of lattice constants. The optimum was reached after 7 iterations; the results are given in Table 11-3.

It is seen that the Ar structure is reproduced perfectly, the KCl structure to 40°/oo, and the lattice energy of Ar to 70°/oo. The restriction on parameters A and B forces K and Cl to be equal in the van der Waals sense, which leads to the inaccuracy in lattice constant and consequently lattice energy of KCl.

11.6.2 Alkanes

As a first test of optimisation on molecular crystals, we optimised a parameter set on the unit cells of ethane[184], pentane[162] and hexane[194], lattice energies[245] of pentane and hexane, and selected conformational and vibrational data of the gaseous molecules used in the optimisations described in chapter 9. As parameter input we used PEF302opt, and let only K_ϕ, $\sqrt{A}$ and $\sqrt{B}$ vary. The resultant set, PEF303opt, is shown in Table 9-1, and its applications to gaseous molecules are discussed in chapters 9 and 10. Results pertaining to the alkane crystals are given in Table 11-4.

In order to see the influence of Coulomb terms, the same parameter set was augmented with atomic charges[223], and optimised as PEF304. The same parameters, plus the two b_o, were allowed to vary. The results are shown in Tables 9-6 and 11-4, and in Figure 9-3. It may be seen from Table 11-4 that the ethane structure is accurate to within 20°/oo, the others to within 40°/oo, and the two lattice energies to within 70°/oo. Calculated unit cell volumes are accurate to 20°/oo. This is just as good as was the case for Ar and KCl.

These encouraging results provoked the further analyses of PEF304 in chapters 9 and 10.

Table 11-4: Alkane crystals in PEF303opt and PEF304opt

		PEF303opt	PEF303opt	PEF304opt	exp.
Summation limit		6 Å	8 Å	6 Å	
Conv. const. (disp.)		0.20	0.17	0.20	
Ethane	A	4.333	4.331	4.302	4.226
	B	5.580	5.580	5.576	5.623
	C	5.730	5.732	5.723	5.845
	β	92.67	92.70	92.34	90.41
	V	138.40	138.40	137.17	138.89
	LE	−25.22	−25.28	−25.50	(−22.00)
Pentane	A	4.287	4.290	4.273	4.10
	B	8.959	8.953	8.952	9.076
	C	14.824	14.870	14.761	14.859
	V	569.35	571.13	564.64	552.93
	LE	−48.63	−49.23	−49.38	−46.4
Hexane	A	4.310	4.308	4.293	4.17
	B	4.546	4.547	4.546	4.70
	C	8.678	8.691	8.639	8.57
	α	97.90	97.83	97.90	96.6
	β	90.61	90.31	90.58	87.2
	γ	102.36	102.26	102.40	105.0
	V	164.43	164.72	163.02	160.88
	LE	−58.23	−59.56	−58.91	−55.2

Only fools and well-trained scientists
keep on wondering why

Sailor: People in love

Some of the questions quoted in the beginning of chapter 3, or just implicit in the text, can now be answered: in what follows, I do my best to guide colleagues who want to get started.

12.1   General guidelines

First of all: analyse your needs. Which types of compounds you want to work on, which properties you are mostly interested in, how accurate results you require, the intended amount of work if you dare predict it.

Secondly: examine your possiblities. Computer and consultant availability, your own or your associates´ programming skills, your budget.

When pondering the above, be aware that you may easily get ideas on new types of substances which have not been parametrized, and you may want to calculate properties which are not available in any or most programs. As to accuracy, remember that high accuracy on conformations does not necessarily cost many parameters but that good vibrational spectra certainly do.

Good ideas are sometimes gotten in awkward places. You may be working at an institution with few or no computer facilities. Then you must drop the idea or go elsewhere. Or you might have all the facilities you can wish for, but lack the skill to use them. Then you are at the mercy of your splendid graduate student.

If you are a newcomer to the field: be consistent in your choice of program, potential energy function and parameter set. Some of them, such as Allinger´s, come as packages; others, such as ours, do not. As a rule, a program written to handle a complex potential energy function can also use a simpler; the opposite is usually not immediately the case, but the program may be easy to modify, as ours is. To use parameters from a complex potential energy function in a simpler one may give worse results than using those devised for the simpler.

All parameter sets available were originally selected, usually from many and different sources, and then fitted, mostly by trial and error, and in a few cases by optimisation, to selected sets of experimental data. Most of this process is indeed subjective, and the originator of a particular parameter set may not have put emphasis on the same details as you would have done. This should not induce you to Bessermachen: do not introduce changes until you are fully acquainted with the set you have, its virtues and shortcomings. Only then can you make sensible and rational changes without impairing the results of the originator´s large and careful work.

If you want to develop your own parameter set, I wish you good luck, but first have a look at sections 3.4.1, 3.4.2, 3.4.7, 3.17.1, 3.17.2, 4.1, and chapter 9. It is hopefully not necessary to emphasize that such work is not a one-person job.

12.2 Packages

After getting all those warnings off my chest, let me recommend you to get hold of a program etc. which will suit you. You are not bound to choose among what is available through Quantum Chemistry Program Exchange[45,p.317] or through commercial channels. If in doubt on how to get what you have selected as the most promising for you purposes, write to the author.

A few notes will bring Burkert and Allinger´s QCPE-table up to date. No. 247, QCFF/PI, see section 3.4.5, was renewed in 1983. Nos. 286, ECEPP, and 361, UNICEPP, were superceded in 1983 by No.454, ECEPP; see section 3.17.2. Williams´ program, see chapter 11, is available from 1984 as No. 481, PCK83. The QCPE Bulletin often brings corrections and hints to the MM family. The announcement in 1984 of No. 476, MMHELP, suggests that the MM programs are maybe not extremely easy to use.

You may be in doubt about which program to order or whom to write to. The choice is of course not strictly objective, but let me stick my neck out and utter some characteristics.

To the best of my knowledge, Allinger´s programs are most like black boxes. They are probably particularly suitable for organic chemists who want to study new synthesis routes. They have been parametrized for many classes of compounds, and it must be expected that both programs and parameter set will be updated.

Far more versatile is CHARMM[35], which maybe is the best all-round system on the market; see also section 3.5 As far as I can judge it, skill in handling large programs is a necessity for successful installation and operation. Unfortunately, optimisation is not included.

Among the other very good programs we find Bartell´s, Boyd´s and Faber´s, of which more is said in sections 3.9, 3.10 and 3.11.

All members of the consistent force field family of programs are available, from QCPE or from the authors; they are described under section 3.4 They are all singular in a way: they allow for optimisation of potential energy function parameters on many types of observable, and are therefore splendid tools in the hands of a skilled worker.

Scheraga´s programs seem to be used less than they deserve, as far as I can judge from literature. For peptide work, the ECEPP, see section 3.17.1, should be as good a candidate as CHARMM or a CFF, provided one can live with stiff bonds and angles. The unconventional form of EPEN, see section 3.17.2, may have scared some; but the idea is so refreshing that it deserves independent use and evaluation.

12.3 Parameter sets

Many people have been hoping for a parameter set which would be globally applicable. Considering the crudeness of even the most elaborate model potential energy functions, that hope will probably never be fulfilled.

Those who might want to put together their own parameter sets should therefore not be totally discouraged by my remarks in the beginning of this chapter. I shall mention here some fortunate compositions of the past; they have undergone revisions and extensions, and more changes and additions will doubtlessly come.

A set of van der Waals functions, mostly, but not exclusively, of the exp-6 type, was proposed by Liquori[156] and soon changed by himself and his associates[51,157], and used by many other authors, also on substances quite different from the peptides Liquori aimed at. The Australians, for example, use them as part of their parameter set for coordination complexes[39], and so do we[186,191], which means on systems quite different from those it was developed for[58,59]. Also Scott and Scheraga developed, very early[240], a Buckingham parameter set which has been used extensively.

For recently developed parameter sets, see chapter 3, but remember that they will be improved also in the future. Among general-purpose parameter sets note also Hopfinger's[112] and a more recent one[252], both with Lennard-Jones functions; the second with stiff bonds and angles.

Many comparisons have been made, most of them dealing only with alkanes or, at most, hydrocarbons. Among those who discuss the basic properties and differences of various potential energy functions I refer particularly to Burkert and Allinger[45,chapter2] and Osawa et al.[198]. In most comparisons, the authors tend to find their own choice of parameters better than what else is available; as an example, see a paper by Zhurkin et al.[288] on nucleic acid structure.

Islam and Neidle have made a comparison[119] between three parameter sets: one of their own composition, Hopfinger's[112] and that of

Stuper et al.[252]. They applied these three to just one molecule with two torsional degrees of fredom, using stiff subunits. Despite rather different details in conformational behaviour, the three sets were equally good in reproducing the crystal conformation of daunomycin.

For good measure, I can also recommend our own sets, especially the newly optimised ones, and those, based on them, that are to come.

Appendix 1   Availability of the Lyngby CFF program

Why not follow the way thyself

Kipling: Kim (Teshoo Lama to Mahbub Ali)

Anybody interested in using our consistent force field program may obtain it from the Danish Data Deposit. Write to

RECKU
Vermundsgade 5
DK-2100 København
Telex: 15069 recku dk

and ask for a copy of DANDOK 0001-84, mentioning, for safety's sake, this book.

You will get a 600´ tape, 9 trk, 6250 bpi, NL, with four files: the source text in FORTRAN compatible with FORTRAN IV and FORTRAN 77, a test input (the cyclohexane example from section A2.3) and the resulting output, in two files. Fixed record lengths are used, the first two files as card images, the third as line printer images, with control characters.

This will cost you dkr 500; you will be billed by RECKU.

The main program and the first subroutine TID contain calls of installation dependent routines. Any programmer can substitute with local versions of TIME (unit used is microsecond), DATE and CLOCK (used by CFFLOG, which returns (ddmmyy, hhmmss)).

I shall endeavour to update the program when important developments occur.

The user manual is found in appendix A2.

I should be very grateful if colleagues who use the program, as received or after having introduced changes, would be kind enough to keep me informed. In return they would receive circular letters mentioning developments at infrequent intervals.

Appendix 2   User Manual

By Birgit and Kjeld Rasmussen

"Why," said the Dodo,
"the best way to explain it is to do it"
(And, as you might like to try the thing
yourself some winter day,
I will tell you how the Dodo managed it.)

Louis Carroll:
Alice's Adventures in Wonderland

The following is a slightly edited version of our Input Manual as it is handed out to students and guests, and sent to optimistic colleagues. Its first edition was written many years ago based on a short concept from Svetozar R. Niketič; very recently Lars-Olof Pietilä has also contributed.

Every time the main author has given a course on the consistent force field, the coauthor has criticised his instructions on how to do it, and advocated a down-to-task approach. The present form, and especially section A2.3, results from this pressure.

Section A2.1 aims specifically at IBM installations; everything else is independent of the actual machine used.

A2.1 Summary of the JCL for CFF under IBM MVS

The programme is operated by two jcl procedures CFF and CFFSTAT.

A2.1.1 Routine operation

CFF is intended for routine use:

```
//A123456  JOB (kkk,NEU,tt,ll),´USER´,REGION=400K
/*ROUTE  PRINT LOCAL
//        EXEC CFF,PAM=thesepar,COR=thesecor,
//             INPUT=thisinput
```

Users outside our own group should, in addition, specify on the EXEC
line:

```
PARLIB=´NEU.A123456.theirpar´,CORLIB=´NEU.A123456.theircor´,
INPUTS=´NEU.A123456.inputlib´
```

Input may also be given in the stream as

```
//CFF.INPUT DD *
```

For estimates of times, lines and region use your own accumulating
experience.

  For preparation of input to ORTEP, MONSTER, PLUTO, MOLVIB and
IBMOL, and to the UNIRAS system, Kjeld Rasmussen must be consulted.
Also for such special activities as drawing of normal coordinates,
statistical analysis of non-bonded interactions, and charting of
pathways in conformational space.

A2.1.2 Program changes

CFF can be used for routine jobs with minor changes in the programme:

```
//        EXEC CFF, etc. ,EMEM=(usually)CORINA
//EDIT.EDIT DD *
```

(control lines for EDITOR)

With this procedure the text editing programme EDITOR is called to operate on the source text of the conformational programme, which the user must have at hand. The manual for EDITOR can be had at NEUCC.

The first control line for EDITOR must be

```
./ SET COLUMN=77/80,NOLIST (or RESULT)
```

The next should invoke selection of the subroutine(s) in which changes are to be made, e.g.:

```
./ SEL ´CHPO´,´NFUO´
```

whereupon follow statements like

```
./ REP or
./ INS or
./ DEL or
./ CHA
```

An alternative is to use a permanent member containing an edited version of one or a few subroutines:

```
//          EXEC CFF,CMEM=(new subroutine(s)), etc.
```

Or you may use line image input for the compiler:

```
//FORT.SYSIN   DD *
```

```
     (new subroutine(s))
```

This is more clumsy to use, as the entire subroutine(s) must be written conforming exactly to the INTEGER and REAL declarations of the programme and to its COMMON structure. This feature is intended for programme development in our group.

Access to a batch editor is essential as soon as any further development of the programme is attempted. Users of IBM machines without access to EDITOR should get it speedily from Weizmann Institute Computer Center, Rehovot, Israel. Other users should approach their own computer centre to obtain similar facilities.

The DD statements, which make the IBM job control language more complicated than Russian and even Hebrew, are included in the jcl procedures and need not be considered by the general user, with the exceptions mentioned above and below.

INPUTS is a permanent line image data set.

The permanent files PARLIB and CORLIB should be unformatted and blocked PO datasets:

```
PARLIB: DCB=(RECFM=VBS,LRECL=2494,BLKSIZE=2498)
CORLIB: DCB=(RECFM=VBS,LRECL=1624,BLKSIZE=1628)
```

Some users may prefer to put parameters in each time they run a job, and refrain from using permanent sets.

In such case a line:

```
//CFF.PARLIB DD UNIT=DISK,DISP=(NEW,DELETE),
//    SPACE=(TRK,(1,1)),LABEL=(,,,IN),DCB=as above
```

should be placed after the other lines mentioned.

The use of a similar temporary dataset for coordinates is possible but impractical.

In routine minimisation and optimisation runs, output from BRACK, CODER and MKLIST will be identical from one job to another. It can be killed by writing ,LOG2=DUMMY on the EXEC line.

In routine runs compiler output may be killed with ,CPRINT=DUMMY. Default compiler output is NOSOURCE,NOMAP. If you want something else, write on the EXEC line: ,COPTION=´......´.

NEWSLETTER is killed with ,NPRINT=DUMMY, input check with ,IPRINT=DUMMY, and EDITOR output with ,EPRINT=DUMMY. Linkage editor output is generally without interest; it is killed with ,LPRINT=DUMMY.

Table A2-1: Organisation of background memory

Reference number (LOG) external	internal	Contents	Produced by subroutines	Used in subroutines
8	IND	Atomic coordinates	BRACK	CONFOR
9	IUD		CONFOR	BUILDZ THDYN
10	IPAM	Energy parameters	NPAR	NPAR
11	ICTR	Control parameters for individual molecules	BRACK	CONFOR VIBRAT BUILDZ THDYN
12	IORD	Packed words	MKLIST	CONFOR VIBRAT BUILDZ
13	IDIP	Calculated dipole moment and charges	CONFOR	BUILDY BUILDZ
14	IBMA	DSX matrix	CONFOR	VIBRAT
15	IBMX	D vector, DSX matrix, DD matrix	CONFOR	BUILDZ
16	ICON	Calculated conformations	CONFOR	BUILDZ BUILDY
17	IEXP	Measured observables	RDEXP	BUILDZ BUILDY
18	IZMA	Z matrix	BUILDZ	ZMATRX
19	IEIV	Eigenvectors	EIGEN	VIBRAT
20	ICOR	Final atomic coordinates	CONFOR	BRACK
21	IFRQ	Calculated frequencies	VIBRAT	BUILDY THDYN
22	IFMA	DD- matrix	CONFOR	BUILDZ
23	INTE	Fractional charges	CONFOR	VIBRAT

24	IEGV	Mass-deweighted eigenvectors	VIBRAT	BUILDZ
27	IORT	Input for ORTEP,PLUTO,MOLVIB, IBMOL,COLOR	TORTEP TPLUTO TMLVIB TIBMOL TCOLOR	
29	ICRY	Symmetry and convergence information for crystal calculations	BRACK	CONFOR
30	ICRT	Interaction table for crystals	CRYSTA	CONFOR
31	ILEN	Lattice energy and lattice energy derivatives	CONFOR	BUILDY BUILDZ
37	IMST	Input for MONSTER	TMONST	

Files IPAM and ICOR are intended to be permanent. Files IORT and IMST will most rationally be kept until they are used by ORTEP, PLUTO, COLOR or MONSTER, or transferred for use by MOLVIB or IBMOL, and then deleted. The other files are usually deleted at the end of a job, but may be kept for special applications by overwriting the jcl.

In optimisation, great masses of (hopefully) uninteresting output from intermediate minimisation can be killed with
,LOG6=DUMMY. Optimisation output appears on LOG3.

A2.1.3 Subprogram statistics

CFFSTAT loads the programme with a monitor FORTSTAT, written by Henrik Dahlin of NEUCC. It counts every call of a subprogramme and measures the cumulated time spent in each. It is intended for use in the development of more effective programme structures.

A2.2 Input to the conformational program

Line form 1 Front page text	FORMAT (18A4)

AMESS   Any message to be printed on the title page of the output. Columns 1 - 72 may contain any valid EBCDIC character. May be blank but a line must be present.

Line form 2 Global control parameters	FORMAT (10I2,F10.3)

Columns:

1 - 2   KJELD   specify the choice of potential energy functions. For usage of parameters KJELD

3 - 4   NIKI   and NIKI see Table A2-2

5 - 6   NBTYP   specifies the form of non-bonded potentials.

                The options are:

                0 --- default value 5 is used
                1 --- Lennard-Jones 9-6 with parameters A, B and e
                2 --- Lennard-Jones 12-6 with parameters A, B and e
                3 --- Lennard-Jones 9-6 with parameters epsilon, r* and e
                4 --- Lennard-Jones 12-6 with parameters epsilon, r* and e
                5 --- Buckingham exp-6 with parameters A, B, C and e

Choice of single atom or atom pair parameters is made automatically from the
parameter inputs; see Table A2-3.

For work on crystals there is the severe
limitation that only Lennard-Jones 12-6
and 9-6 functions of the A,B form are
possible, and only with single atom
parameters.

7 - 8   NMOLEC = number of molecules.

9 - 10   ISAVE   save cartesians from minimisation in
                  a file if ISAVE > 0.

11 - 12   IPAR   specifies where the energy parameters are
                  to be taken from.

                  The IPAR codes are as follows:

                  0 --- read energy parameters from   lines
                  1 --- read energy parameters from   a file
                  2 --- read energy parameters from   a file
                        and update from lines.

13 - 14   NOPTIM = number of least square optimisations
                  of energy parameters. If NOPTIM = 0 all
                  other calculations than optimisation may
                  be done. If NOPTIM = -1 only topological
                  codings are done, but with full print-
                  out of lists of all interactions.

                  Note this particular option which should
                  be used whenever the study of a new mole-
                  cule is begun.

21 - 30   XLAMBDA Lagrange multiplier in repeated optimisa-
                  tion.

Parameters on this line are common to the entire computation.

Line form 3 Energy parameters   FORMAT (7A1,3X,6F10.4,I2)

This (and the following lines) specify the energy parameters for
different types of interactions. There are as many lines as there
are types of interactions. These lines should be present only if
IPAR = 0 and NOPTIM > 0 in line no. 2.
   Many users have overlooked the previous statement.

Columns:

1 - 7   Symbolic representation of interaction type. See
          Table A2-3.

Table A2-2: Usage of parameters KJELD and NIKI

Subroutine	Parameter value	Performance
MKLIST	KJELD < 0	1,2 and 1,3 interactions are excluded from the list of non-bonded interactions.
	KJELD > 0	1,2 interactions are excluded from the list of non-bonded interactions, whereas the 1,3 and higher interactions are included.

NIKI controls the treatment of torsions. It is possible to include all torsional angles, one torsional angle, or to exclude torsional interactions at metal-ligand or at other chain bonds.

		Metal-ligand bonds	All other chain bonds
	NIKI < 0	None	One phi angle
	NIKI = 0	One phi angle	One phi angle
	NIKI > 0	One phi angle	All phi angles

Subroutine	Parameter value	Performance
MOLEC	KJELD <-1	Torsional and Urey-Bradley terms are included in the computation of total energy.
	KJELD >-1	Urey-Bradley terms are omitted.
	KJELD > 1	Torsional and Urey-Bradley terms are omitted.
EBOND and BFUNC	KJELD < 0	All bond contributions are calculated with harmonic functions.
	KJELD = 1	All bond contributions are calculated with Morse functions.
	KJELD > 1	All bond contributions are calculated with inverse power functions.
EKHI	NIKI = 0	Only end and apex atoms matter in out-of-plane bending.

| 8 - 10 | Blank. |

Table A2-3: Coding of energy parameters

Type of	Code	Energy parameters				
interaction	COLUMNS: 1234567	COLUMNS: 11-20	21-30	31-40	41-50	51-60
Bond end	H-				(not implemented)	
Bond	N-H	K(b)	b(0)		(harmonic)	
		D(e)	a	b(e)	(Morse)	
		A	B	C	(inverse power)	
Valence angle	C-C-N	K(θ)	θ(0)	F	F$^\prime$	d(0)
Torsional angle	M-N-C-C	K(ϕ)	n			
Out of plane	Q(K-C-C	K(χ)	χ(0)			
Atom	H--	A	B	e	(Lennard-Jones plus Coulomb)	
Atom pair	H---H	A	B	C	e (Buckingham or Lennard-Jones)	
		eps	r*			
Fractional charge	H.	e				

F, F$^\prime$ and d(0) = Urey-Bradley parameters
K(ϕ) = torsional potential barrier
n   = multiplicity of torsional barrier
Buckingham parameters must always be specified for atom pairs. For
Lennard-Jones parameters: Examples given; check with formulae.

Units:

K(b) kcalmol(-1)Å(-2)   b(0) Å

D(e) kcalmol(-1)   a   Å(-1)   b(e) Å

K(θ) kcalmol(-1)rad(-2) θ(0) rad

K(χ) kcalmol(-1)rad(-2) χ(0) rad

F   kcalmol(-1)Å(-2)   F$^\prime$ kcalmol(-1)Å(-1)   d(0) Å

K(ϕ) kcalmol(-1)   n

A   (kcalmol(-1)Å(9))(1/2) B   (kcalmol(-1)Å(6))(1/2)

```
A    kcalmol(-1)*10(-4)      B     A(-1)

C    kcalmol(-1)A(6)         e     elementary charge

eps  kcalmol(-1)             r*    A
```

Table A2-4: Atomic symbols[190]

Symbol	Atom
H	Hydrogen
D	Deuterium
X	Halogen
Q	Double bonded oxygen
O	Oxygen
A	non-tetrahedral nitrogen
K	Double bonded carbon
C	Tetrahedral carbon
N	Tetrahedral nitrogen
M	Octahedral, tetrahedral or square-planar metal
T	Sulfur (not fully implemented)
P	Phosphorus (not fully implemented)

11 - 20	Values of energy parameters.
21 - 30	ditto
.....	
.....	
61 - 70	ditto
71 - 72	Delimiter: nonzero integer to indicate the last of parameter lines.

There is a possibility of assigning exact values of pi, 2*pi/3, the tetrahedral angle or pi/2 to parameters θ(0). It is done by writing 180., 120., 109. or 90. as the parameter value in stead of the value in radians.

The contents of the next lines depend on the values coded for various control parameters on line no. 2.

If IPAR = 2 (on line no. 2) the program expects to get information on updating of energy parameters which exist on file IPAM. The following two line forms are used for updating of parameters:

Line form 4 Updating of energy parameters   FORMAT (36I2)

Columns:

1 - 2   NOPT = number of parameters to be updated.

Line form 5   FORMAT (5(I3,F11.4))

Columns:

1 - 3   LSTOPT(1) = list number of first parameter to be
                         updated.

4 - 14   POPT (1) = updated value.

15 - 17   LSTOPT(2) = list number of second parameter to be
                         updated.

18 - 28   POPT (2) = updated value.

.....   etc. up to column 70. Should be continued on the
             next line(s) if necessary, up to LSTOPT(NOPT) and
             POPT(NOPT).

Line form 6 Molecule   FORMAT (9A8)

ANAME   Name of the molecule. Up to 72 characters in
             columns 1 - 72.

Line form 7 Formula   FORMAT (72A1)

SYMB   Chemical formula in columns 1 - 72. Can be con-
             tinued on one or more lines (columns 1 - 72).
             Maximum number of characters is 300.

Table A2-4 shows the allowed atomic symbols and their interpretation by the programmes. Digits 1-6 are interpreted as subscripts on sideatoms. Sidechains are enclosed in parantheses, as is the entire formula. R and S before chain atom specify the chirality, before M they mean DELTA and LAMBDA, respectively. Rings are closed with pairs of special symbols (e.g. ; & * , . : A etc.) placed before two chain atoms bound to each other.

A molecular complex is enclosed in an extra set of parantheses. Ionic charges are coded with + and - ; this should be used when Coulomb terms are present in the potential energy expression.

The asymmetric unit of a molecular or ionic crystal is enclosed in an extra set of parentheses as for a molecular complex.

It is imperative to check the coding whenever operations on a new molecule are begun.

Examples:
----
n-butane   (CH3CH2CH2CH3)

cyclohexane   (.CH2CH2CH2CH2CH2.CH2)

1,1,3,3-   (CHX2CH2CHX2)
-tetrachloropropane

DELTA-   (,&;RM(NH2CH2CH2;NH2)(NH2CH2CH2&NH2)(NH2CH2CH2,N
(Co(en)(3))   H2))

Water tetramer ((OH2)4)

Outer-sphere   ((:&;RM(NH2CH2CH2;NH2)(NH2CH2CH2&NH2)(NH2CH2CH2:
complex   NH2))(+3)(PQ4)(-3))

The following line forms contain information on each of the molecules to be studied. There should be NMOLEC sets of these lines.

Line form 8 Control parameters   FORMAT (36I2)
             for individual molecules

Columns:

  1 - 2   ICART   specifies the type of input of cartesian
                         atomic coordinates. The ICART codes are
                         as follows:

                         1 --- cartesians for all the atoms are
                               read from the lines following
                               this.

>>> 2 --- cartesians of chain atoms are read from the following lines. The rest e.g. hydrogen atom coordinates, are generated by the program. WARNING: This works only in certain cases.
>>> 3 --- coordinates are read from file ICOR.
>>> 4 --- all coordinates are generated by the program. WARNING: This works only in certain cases.

3 - 4	IOR	specifies the number of the atom to be placed in the origin of the coordinate system.

5 - 6	IXX	specifies the number of the atom to be placed on the +X axis.

7 - 8	IXY	specifies the number of the atom to be placed in the XY plane.

9 - 10	NO = 1	calls transformation of coordinates as above.
	= 2	calls transformation to coordinate system defined by the principal moments of inertia.
	= 0	Transformation is not performed.

11 - 12	METAL	specifies the type of the central metal atom in coordination compounds by its number in the Periodic System.

13 - 14	NCHARG	specifies the total ionic charge.

15 - 16	NSIGMA	specifies the symmetry number of the molecule.

17 - 18	NSTEEP	specifies the maximum number of iterations in steepest-descent minimisation.

19 - 20	NDAVID	specifies the maximum number of iterations in Davidon-Fletcher-Powell minimisation.

21 - 22	NEWTON	specifies the maximum number of iterations in modified Newton minimisation.

If NSTEEP, NDAVID or NEWTON=99 it is changed to 201.

In addition, the following logic operates:

If the sum of the three minimisation parameters is zero, analytical and numerical values of all first and second derivatives are compared and errors printed. More than two per cent errors will prevent further calculation on the molecule.
If the sum is -1, analytical and numerical first derivatives are printed.
If it is -2 or less, analytical and numerical first and second derivatives are printed.
These functions require much CPU time.

23 - 24   ICONF<0   allows for calculation of initial conformation and coordinate transformation, but not for minimisation.
    =0   bypasses all conformational calculations.
    >0   allows for derivative checks, initial conformation and minimisation.

25 - 26   IFREQ=0   bypasses vibrational calculations.
    =1   calls for frequencies.
    =2   calls for vibrational frequencies and cartesian displacement coordinates.
    =3   calls for vibrational frequencies and changes in normal coordinates in terms of internal coordinates.
    =4   gives a sorting of internal displacement coordinates for assignment purposes.

27 - 28   INTEN>0   If charge parameters are used, infrared intensities may be calculated.
    =1   IR intensities.
    =2   IR circular dichroism (rotatory strength/dipole strength).
    =3   DS, RS, CD, oscillator strength.
    If IFREQ>1, INTEN=3 works as INTEN=2.

29 - 30   ITDYN>0   calls thermodynamic functions.
    =1   vibrational terms at 300K
    =2   vibrational terms at 273 and 300K
    =3   vibrational terms at 100, 200 and 300K
    =4   vibrational terms at 100, 200, 300 and 375K
    If 10 is added to ITDYN, the programme gives also translational and rotational contributions, and rotational constants and principal moments of inertia.

31 - 32   NPHIIN   number of torsional angles to be specified for REDUCE. To be used only for ICART = 4.

33 - 34 NANGIN number of extra angles defined by four
          atoms to be read by MKLIST.

35 - 36 IORTEP>0 prepares input for programme ORTEP on
          file IORT.

37 - 38 IMONST=0 prepares input for programme MONSTER
          on file IMST.
         >0 all bonds to be drawn.
         <0 sideatom bonds to H and D are omitted.

39 - 40 ICRYST>0 formula is interpreted as a unit cell.
         =0 formula is interpreted as a molecular
         complex.

41 - 42 IPLUTO=0 prepares input for programme PLUTO on
          file IORT.
         =1 Dreiding-type models
         =2 ball-and-spoke models
         =3 CPK-type models
         <0 as above, H and D left out

43 - 44 MOLVIB=0 prepares input for programme MOLVIB on
          file IORT.
          Only C, H, X are implemented.
        >0 all interactions.
       =-1 no torsions.
       <-1 no torsions, no non-bonded interactions.

45 - 46 NIBMOL>0 prepares input for programme IBMOL on
          file IORT.
          NIBMOL is used as the number of the
          atomic orbital basis set.

47 - 48 NSTATW>0 The statistical weight NSTATW of this
          particular conformer is taken into
          account when free enthalpy is calculated

49 - 50 ICOLOR=0 prepares a programme for the UNIRAS
          raster plot system written on file IORT.
         =1 perspectivic
         =2 stereo
         =3 perspectivic; Dreiding overlayed
         =4 stereo; Dreiding overlayed
       10 added Versatec simulation

51 - 52 NDNGIN Number of torsional angles as defined by
          the program to be deleted. List numbers
          of these are read by MKLIST.

According to the ICART code specified on the previous line we may
read either all the coordinates or the coordinate of chain atoms
with the following format, or none.

If ICRYST≠0 the following lines, 9 to 19, are added.

Line form 9 Lattice   FORMAT (6F12.9)

The lattice constants a, b, c, alpha, beta, gamma, and the position
of the asymmetric unit in fractional coordinates. If a lattice constant is given a negative value, it is constrained to the given
(positive) value during the minimisation. Lattice angles are given
in degrees. Even when the coordinates, lattice constants and position of the asymmetric unit are read from a file (ICART=3) these
lines must be present. In this case only the signs are important.

Line form 10 Summation   FORMAT (3F10.4)

The summation limit in Å.

Line form 11 Convergence   FORMAT (3F10.4)

The convergence constants for coulombic, attractive and repulsive
interactions.

Line form 12 Site symmetry   FORMAT (24I3)

Columns:

  1 - 3   NODA   Number of site symmetry operations (<4).

  4 - 6   NINDEP   Number of independent atoms.

  7 - 9   NIALS(1) NIALS(I) is the number of atoms
                        on the site symmetry element I.
 10 - 12   NIALS(2)

 13 - 15   NIALS(3)

 16 - 18   NIALS(4)

If NODA = 0, lines 13 - 18 are omitted.

Line form 13 Site symmetry operations   FORMAT (3F12.9)

Columns:

  1 - 12   RPS(I,1) Site symmetry operations are given row
                        by row.
 13 - 24   RPS(I,2)

 25 - 36   RPS(I,3)

To be repeated, I = 1 to NODA, 3 lines, each with 3 numbers, for each I.

Line form 14 Atoms on site symmetry elements   FORMAT (24I3)

Columns:

1 - 3   IALS(1)   Numbers of atoms on site symmetry
                     elements.
4 - 6   IALS(2)   First atoms on symmetry element 1,
                     then on symmetry element 2, and so on.
7 - 9   IALS(3)

....

Line form 15 Coordinate indices   FORMAT (24I3)

Columns:

1 - 3   KAKO(1)   Coordinate indices for each site
                     symmetry element.
4 - 6   KAKO(2)   KAKO(I) = 0 or 1.

7 - 9   KAKO(3)   If KAKO(3*(J-1)+I) = 0 the atoms on the
                     symmetry element J are not allowed to
10 - 12   KAKO(4)   move in the direction I (x,y or z).

....

Line form 16 Numbers of generated atoms   FORMAT (24I3)

Columns:

1 - 3   NGAT(1)   NGAT(I) is the number of generated atoms
                     for the site symmetry operation I.
4 - 6   NGAT(2)

7 - 9   NGAT(3)

10 - 12   NGAT(4)

Line form 17 Independent atoms   FORMAT (24I3)

Columns:

1 - 3   INDEP(1) Numbers of independent atoms.

4 - 6   INDEP(2)

....

```
Line form 18 Generation of atoms            FORMAT (24I3)
```

Columns:

```
 1 - 3      IDEP(1)   Generation  of  the  atoms for each site
                      symmetry operation;
 4 - 6      IDEP(2)   first for operation 1, and so on.

 7 - 9      IDEP(3)   IDEP(2*I+1) is generated from
 .....      .....     IDEP(2*I+2).
```

```
Line form 19 Symmetry                       FORMAT (3F12.9)
```

The symmetry operations and their positions. The transformation
matrix is given in cartesian coordinates, one row per line, and the
translation and position vectors are given in fractional coordinates. In all ICRYST-1 symmetry operations.

```
Line form 20 Cartesians                     FORMAT (3F10.4)
```

Columns:

```
 1 - 10

11 - 20      X, Y, and  Z  coordinates  of  all the  atoms (if
             ICART=1) or of chain atoms (if ICART=2) per line.

21 - 30
```

```
Line form 21 Reference                      FORMAT (9A8)
```

```
REF          Reference  for cartesians.  Any text  in  columns
             1 - 72.  May be blank but  a line must be present
             if  the coordinates are read from input  (i.e. if
             ICART=1 or 2).
```

If ICART is different from 4, line 22 is omitted. If NPHIIN=0 it is
also omitted.

```
Line form 22                                FORMAT (12F6.1)
```

Columns:

```
 1 - 6       PHIVAL(1) = value of the first torsional angle in
                         degrees.

 7 - 12      PHIVAL(2) = value of  the second torsional angles
             in degrees.

 .....       etc.    To be continued on  the next line if ne-
             cessary, up to PHIVAL(NPHIIN).
```

If NDNGIN = 0 line 23 is omitted.

Line form 23 Unwanted torsions   FORMAT (24I3)

Columns:

  1 - 3   NDELPH(1) Numbers of the unwanted torsional
                         angles to be deleted.
  4 - 6   NDELPH(2)

  7 - 9   NDELPH(3)

  ....

If NANGIN = 0 line 24 is omitted.

Line form 24 Extra torsions   FORMAT (36I2)

Columns:

  1 - 2   IP(1) Atom indices defining first of the extra
                      torsional angles.
  3 - 4   JP(1)

  5 - 6   KP(1)

  7 - 8   LP(1)

  9 - 10   IQ(1) Bond order of central bond.

 11 - 12   IO(1) Bond code of central bond.

               One line per angle.

    If NMOLEC is greater than 1 (on line no. 2) the sequence of line
forms 6 - 12 is repeated for every other molecule.
    If NOPTIM is greater than 0 (on line no. 2) the following in-
formation on the parameters to be optimised should be coded:

Line form 25 Optimisation of energy parameters   FORMAT (36I2)

Columns:

  1 - 2   NOPT·= number of parameters to be optimised.
                     Max. 20.

  3 - 4   NSTR = number of constraints between parameters.

Line form 26 Optimise   FORMAT (5(I3,F11.4))

Columns:

1 - 3   LSTOPT(1) = list number of the first parameter to
                        be optimised.

4 - 14   Blank.

15 - 17   LSTOPT(2) = list number of second parameter to be
                        optimised.

18 - 28   Blank.

.....   etc. up to column 70. May be continued if ne-
             cessary up to LSTOPT(NOPT).

    Be aware that the forms of lines 24 and 25 are identical to those
of lines 4 and 5, respectively, except that the values of parameters
are not written if the parameters are to be optimised.

Line form 27 Constrain   FORMAT (24I3)

Columns:

1 - 3   LSTSTR(1) = listnumber of   the   parameter con-
                        strained equal to that given next.

4 - 6   LSTSTR(2) = listnumber of a   parameter to be
                        optimised.

7 - 9   LSTSTR(3)

....
This line is present only if NSTR > 0 in line 25.
    If NOPTIM > 0, information on data to optimise on is coded as
follows:

Line form 28 Experimental data FORMAT (3(I4,A4,F8.5,F7.5),I3)

Columns:

1 - 4   LSTINT   = list number of quantity.

5 - 8   TYPINT   = type of quantity (not used in CFF,
                        may be blank)

9 - 16   VALINT   = experimental value.

17 - 23   UNCINT   = experimental uncertainty.

24 - 27	LSTINT	= ditto
28 - 31		etc.
32 - 39		etc.
40 - 46		etc.

. . . . .

| 70 - 72 | NOK | = delimiter; non-zero integer indica- |
| | | ting last line for present molecule. |

This group of lines is repeated for each type of observable (conformation and lattice constants, lattice energy, dipole moment,
frequencies, in this order) and again for each molecule. A line must
be present for each type for each molecule even if it contains only
NOK. For lattice energy LSTINT = any positive integer.

If UNCINT is not given, it is set to a standard value. This is
rather practical for frequencies.

If frequencies are not specified for some molecule, IFREQ (line
form 8) must be 0 for that molecule. ICONF must be 1 in any case.

This information is read from file EXPIN; see A2.1.

A2.2.1 Cautioning

Optimisation of Urey-Bradley parameters F,F′ and d(0) is not
implemented.

Optimisation on charges is not implemented in the standard version. It may be wanted when optimising a PEF which includes charge
parameters. In this case there is one more type of observables for
each molecule; look into subroutine RDEXP or consult us. As for the
standard version, all lines must be there.

A dipole moment not known, or known to be zero by symmetry, is
specified with a line containing only NOK. (Symmetry must emerge in
a more fundamental way than by imposing zero dipole moment.)

Table A2-5: Checklist of READ statements

```
-----------------------------------------------------------------
Program Line Read                              Format
        Form
-----------------------------------------------------------------
MAIN     1   AMESS                             18A4

         2   KJELD, NIKI, NBTYP, NMOLEC,       10I2,F10.3
             ISAVE, IPAR, NOPTIM,,,,XLAMBDA
-----------------------------------------------------------------
NPAR     3   (SYMB(I),I=1,7),(PX(J),J=1,6),    7A1,3X,6F10.4,I2
             NEXT

         4   NOPT                              36I2

         5   (LSTOPT(I),POPT(I),I=1,NOPT)      5(I3,F11.4)
-----------------------------------------------------------------
BRACK    6   ANAME                             9A8

         7   SYM(I)   (1<I<300)                72A1

         8   ICART, IOR, IXX, IXY, NO,         36I2
             METAL, NCHARG, NSIGMA, NSTEEP,
             NDAVID, NEWTON, ICONF, IFREQ,
             INTEN,ITDYN,NPHIIN,NANGIN,IORTEP,
             IMONST,ICRYST,IPLUTO,MOLVIB,
             IBMOL,NSTATW,ICOLOR,NDNGIN

         9   a,b,c,alpha,beta,gamma,           6F12.9
             position of the asym.unit

        10   ELIM                              3F10.4

        11   FKC,FKD,FK                        3F10.4

        12   NODA,NINDEP,(NIALS(I), I=1,4)     24I3

        13   (RPS(J,I), I=1,3)                 3F12.9

        14   IALS(I)                           24I3

        15   (KAKO(I), I=1,12)                 24I3

        16   (NGAT(I), I=1,4)                  24I3

        17   INDEP(I)                          24I3

        18   IDEP(I)                           24I3

        19   XCRY(I)                           3F12.9

        20   X(I)     (1<I<201)                3F10.4
```

	21	REF	9A8
REDUCE	22	(PHIVAL(I),I=1,NPHIIN)	12F6.1
MKLIST	23	(NDELPH(I), I=1,NDNGIN)	24I3
MKLIST	24	(IP(I),JP(I),KP(I),LP(I),IQ(I), IO(I),I=1,NANGIN)	36I2
NPAR	25	NOPT,NSTR	36I2
	26	(LSTOPT(I),blank,I=1,NOPT)	5(I3,F11.4)
	27	(LISTSTR(I),I=1,2*NSTR)	24I3
RDEXP	28	(LSTINT(I),TYPINT(I),VALINT(I), UNCINT(I),I=1,3),NOK	3(I4,A4,F8.5, F7.5),I3

Similarly for charges, if they cannot be estimated from photoelectron spectra or ab initio calculations. (Do not use estimates from CNDO, PCILO, ...; they are usually unreasonably low for a monopole model of charge distribution.)

Users are strongly advised to acquire much experience in minimisation and some in normal coordinate analysis before adventuring into optimisation.

A2.3 An example: cyclohexane

This input is recommended as a standard test both of new programme versions and of new users. Do examine the various sections of
the resulting output carefully.

Line 1

    DEMONSTRATION RUN :   FULL CODINGS

Line 2

-1 0 4 1 0 0-1   All interactions, one torsion
                                per bond, L-J functions with eps
                                and r*, one molecule,,,check of
                                coding only.

Line 6

            CYCLO-HEXANE

Line 7

(&CH2CH2CH2CH2CH2&CH2)

Line 8

 4   Build geometry.

Line 1

  DEMONSTRATION RUN : MINIMISATION, NORMAL COORDINATES,
                       THERMODYNAMICS

                                Note that you may run more than
                                one CFF calculation in the same
                                EXEC, but that you have only
                                one choice of parameters and
                                coordinates. (Do not attempt to
                                run more than one EXEC CFF in a
                                job.)

Line 2

-1 0 4 1 1 0 0   ,,,,save   coordinates   after
                                minimisation, read parameters
                                from the following lines.

Line 3

O–C	720.	1.410	Bond parameters.
C–C	510.	1.509	
O–H	1070.	0.955	
C–H	670.	1.093	
O–C–O	60.	109.	Angle parameters.
O–C–C	100.	109.	109. gives the tetrahedral angle to machine precision; this also
C–O–C	70.	1.80	works for 90., 120., 180.
C–C–C	50.	109.	
O–C–H	100.	109.	
C–O–H	80.	1.80	This is how it should be for other values: use radians.
C–C–H	71.	109.	Urey–Bradley parameters (three in number) are put on the same
H–C–H	75.	109.	line; we seldomly use them.
C–O–C–C	0.001	3.	Torsion parameters. n = –2. for double bonds etc.
C–C–C–C	0.001	3.	
O–––O	0.300	3.00	Non-bonded parameters for pair of atoms.
O–––C	0.300	3.05	
C–––C	0.100	3.50	
O–––H	0.100	2.95	
C–––H	0.100	3.15	
H–––H	0.300	2.75	
O.	–0.400		Charge parameters.
C.	–0.050		
H.	0.125		

1

Remember a number in col 71 – 72 to show there are no more para-meters.

Parameters are PEF400 from ref. 172.

Line 6

CYCLO-HEXANE

Line 7

(&CH2CH2CH2CH2CH2&CH2)

Line 8

 4 4 7 1 1   0 650 020 1 4 111 5   1

> Build geometry, reference coor-
> dinate system: atom 4 in origo,
> atom 7 on x-axis, atom 1 in xy-
> -plane, do the transformation
> after minimisation,, charge zero,
> symmetry number six, fifty steep-
> est descents, no davidons, up to
> twenty newtons, do this, print
> normal coordinates in sorted in-
> ternal displacements, with IR in-
> tensity (only when you have
> charge parameters), do thermody-
> namics for gas at 300K, put in
> five torsions to start with,...,
> statistical weight one (trivial
> here, inportant when more than
> one conformer or isomer are pre-
> sent).

Line 11

   60. -60.   60. -60.   60.

A2.4 A second example: the ethane crystal

This input was used in checking our crystal version and is the standard for this purpose in the present development phase.

Line 1

 + ETHANE + PEF303 +

Line 2

-1 0 2 1 1 0 0   All interactions, one torsion per bond, L-J functions with A and B, one "molecule", save co-ordinates, read parameters from the following lines.

Line 3

C-C   720.   1.52

C-H   720.   1.09

C-C-C   143.9   1.911

C-C-H   93.5   1.911

H-C-H   74.8   1.911

C-C-C-C   0.001   3.

C--   573.6   18.95   Non-bonded parameters for single atoms.

H--   131.7   7.046   1

Parameters are from Pietilä and Rasmussen [207].

Line 6

+++++++ ETHANE , CRYSTAL TEST ++++++++

Line 7

((CH3CH3))   (( )) mean molecular complex or asymmetric unit in crystal.

Line 8

 1   15 1   2

Read geometry ,,,,,,,,,,, up to 15 newtons, do this ,,,,,,,,,,, two asymmetric units per unit cell.

Line 9

4.2286	5.4717	5.6072	-90.	92.51	-90.

0.   0.   0.

Lattice constamts on first line, note constraint on alfa and gamma; position of asymmetric unit on next.

Line 10

6.0

Summation limit 6 Å.

Line 11

0.   0.20   0.

Convergence constants. Here zero for the repulsive part, which saves time and can hardly be noticed.

Line 12

1 4 0

One site symmetry operation (NODA=1), four independent atoms, no atoms on the symmetry element.

Line 13

-1.0	0.	0.
0.	-1.0	0.
0.	0.	-1.0

The site symmetry operation is an inversion.

Line 14

0

No atoms on the symmetry element,

Line 15

0

This line has no influence, as there are no atoms on the site symmetry element.

Line 16

4

Site symmetry operation generates four atoms.

Line 17

1 2 3 4

Numbers of independent atoms.

Line   18

5   1   6   2   7   3   8   4   Atom number 5 is generated from
                                       atom 1, 6 from 2, 7 from 3, 8
                                       from 4.

Line   19

−1.0   0.0   0.0   First three lines: a C_2
                                       operation around the y axis;
 0.0   1.0   0.0

 0.0   0.0   −1.0

 0.0   0.5   0.0   fourth   line: translation
                                      along   y;
 0.25   0.0   0.25   fifth line:
                                      position of the C_2s.

Line   20

−0.173052   0.519995 −0.531312   Cartesian   coordinates,   one
                                      line   per   atom   in   the
 0.264926   0.239329 −1.489197   asymmetric unit.

 0.215037   1.495455 −0.241187

−1.254207   0.592673 −0.644755

 0.173052 −0.519995   0.531312

−0.264926 −0.239329   1.489197

−0.215037 −1.495455   0.241187

 1.254207 −0.592673   0.644755

Line   21

++ STRUCTURE: FROM CALCULATION 2 (PEF303) ++
                        From an earlier run.

Appendix 3   The system of programs

A description of the Lyngby version of the consistent force field programming system was given in chapter 2 of ref. 190. Little has changed since then, apart from the addition and changes of a few subroutines and small utility programs. This may reflect either foresight of the former authors or lack of imagination of the present. Be that as it may, I feel no urge to more or less copy what was written before. I want just to give a short exposition of the system as it is at the time of writing.

A3.1   The CFF system

The consistent force field program, including all subroutines, is a bit over 11000 FORTRAN lines long, and contains some 100 subprograms. It is installed on an IBM 3081 to run under MVS and can, with few changes, run on AMDAHL, CRAY, VAX and UNIVAC machines. It will work just as well on CDC machines, only more changes are necessary. I know that the program has been made to run on a HITACHI machine, but I have no data.

On the Lyngby installation, the program is operated with a JCL procedure CFF, whose structure is shown in Figure A3-1, and which is reproduced in A4.

For minor jobs a corresponding command list runs under TSO. It is reproduced in A5. The procedure is very versatile, as it is possible to specify in the job-setup permanent files for general input, parameters, coordinates and input of observed data for optimisation. There is also a possibility of substituting one or more subprograms with others having the same names, taken from a source library. One may also make changes in the program, which are not kept permanently. This is done with a batch editor which, incidentally, is good for many other purposes. EDITOR comes from the Weizmann Insti-

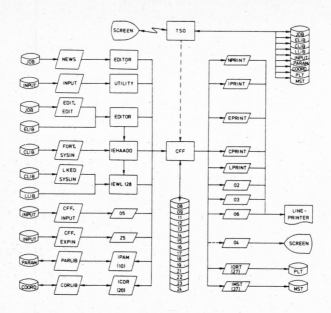

Figure A3-1: Structure of the CFF JCL procedure

tute Computer Center.

A3.2 Interfaces

Outprint from CFF is directed to three files, units 02, 03 and 06. In this way it is possible to kill unneeded output, for instance from the topological parts (unit 02), which is unchanged in routine runs, or from minimisation (unit 06), which may run into several hundred pages during optimisation, or from optimisation (unit 03) when this is not performed. CFF uses 16 scratch files. Four permanent files are used for output: unit 10 for parameters and unit 20 for coordinates, and units 27 and 37 for card image output, which can be used as input to other programs and systems.

In this way we have, quite routinely, interfaces to vector plotting (ORTEP, PLUTO and MONSTER), raster plotting (COLOR and UNIRAS), conventional normal coordinate analysis with optimisation (MOLVIB), and ab initio calculations (IBMOL-6 and GAUSSIAN-70). ORTEP, PLUTO, IBMOL-6 and GAUSSIAN-70 are so well-known that they probably need no more mention here; MONSTER, which is the property of Mr. Per Jacobi, Brede 50, DK-2800 Lyngby, is essentially a program for viewing figures made up of lines. It is intended for use by people unskilled in programming, it has an easy input language, and it is very fast. We use it for stereo viewing of Dreiding-type models of conformers and of amplitudes in normal coordinates [214]. COLOR is a raster system which supports exclusively a Hertz and an Applicon plotter at the University of Lund. UNIRAS is a new raster system, a property of European Software Contractors. It is independent of plotter devices, and we have drivers for Applicon, Hertz, Versatec, Paper Tiger, PRISM, Tektronix and others. MOLVIB, which is the property of Dr. Tom Sundius, Accelerator Laboratory, University of Helsinki, Brobergsterrassen 20 M, SF - 00170 Helsinki, is a very versatile program for conventional normal coordinate analysis. It allows for different types of force field, and for several optimisation algorithms. The input language is particularly friendly.

A3.3 Utilities

We have developed many small programs for copying and listing of stored parameters, for plotting of potential energy functions, for conversion of crystal structure data, and for many sorts of operation on coordinate data sets. In addition to copying to and from tape, we may mention shaking of a molecule, re-orientation of its coordinate system, selection from a set, and substitution by any group of atoms whose coordinates are known.

More programs are steadily being added to the utilities as well as to the CFF system.

Appendix 4   JCL procedure CFF

The procedure most often used on our IBM installation is reproduced
below without changes.

```
//*A108001   USER= R A S M U S S E N   TLF= 3334 & 3368
//*A108001 ,TSO ,´ONLINE SESSION        ´,TUESDAY   26.07.83 , 12.43.
//CFF     PROC CLIB=´A108001.SOURCE´,LLIB=´A108001.LINK´,
//             ELIB=´A108001.SOURCE´,EMEM=DUMMY,EDIT=DUMMY,
//             PARLIB=´A108001.PARAM´,CORLIB=´A108001.KOORD´,
//             INPUTS=´A108001.INPUT´,INPUT=DUMMY,EXPIN=DUMMY,
//             NEWSLIB=´A108001.JOB´,NEWS=NEWS,
//             PAM=RESERVE,COR=RESERVE,ORT=DUM,MST=DUM,
//             COPTION=´NOSOURCE,NOMAP´,CMEM=DUMMY,FORTRAN=IEKAA00,
//             LMEM=NADIA0,LOPTION=´NOLIST,NOMAP´,
//             IPRINT=´SYSOUT=A´,NPRINT=´SYSOUT=A´,
//             EPRINT=´SYSOUT=A´,CPRINT=´SYSOUT=A´,
//             LPRINT=´SYSOUT=A´,LOG3=´SYSOUT=A´,
//             LOG2=´SYSOUT=A´,LOG6=´SYSOUT=A´
//*           WRITTEN ON 29 MAR 76 BY KJELD
//*       NEWS
//NEWS    EXEC PGM=EDITOR,PARM=SS
//STEPLIB   DD DSN=A108001.LINK,DISP=SHR
//SYSPRINT  DD DUMMY
//SYSUT1    DD DSN=&NEWSLIB(&NEWS),DISP=SHR
//SYSUT2    DD DSN=&&NPRINT,DISP=(NEW,PASS),UNIT=DISK,
//          SPACE=(TRK,(1,3)),DCB=(RECFM=FB,LRECL=80,BLKSIZE=2000)
//SCRATCH   DD UNIT=SYSDA,SPACE=(TRK,(1,3))
//SYSIN     DD DSN=&NEWSLIB(NEWSOUT),DISP=SHR
//NPRINT  EXEC PGM=UTILITY,PARM=(1,NO,NO,NO,NO)
//STEPLIB   DD DSN=SYS2.LINKLIB,DISP=SHR
//SYSPRINT  DD &NPRINT
//IN        DD DSN=&&NPRINT,DISP=(OLD,DELETE)
//*       INPUT
//INPUT   EXEC PGM=UTILITY,PARM=(1,NO,NO,NO,NO)
```

```
//STEPLIB   DD DSN=SYS2.LINKLIB,DISP=SHR
//SYSPRINT  DD &IPRINT
//IN        DD DSN=&INPUTS(DUAL),DISP=SHR
//*       EDIT
//EDIT    EXEC PGM=EDITOR,PARM=SS
//STEPLIB   DD DSN=A108001.LINK,DISP=SHR
//SYSPRINT  DD &EPRINT
//SYSUT1    DD DSN=&ELIB(&EMEM),DISP=SHR,LABEL=(,,,IN)
//SYSUT2    DD UNIT=SYSDA,SPACE=(TRK,(1,3)),DISP=(NEW,PASS),
//            DSN=&TRANSFER,DCB=(RECFM=FB,LRECL=80,BLKSIZE=3120)
//SCRATCH   DD UNIT=SYSDA,SPACE=(CYL,(3,1))
//SYSIN     DD DDNAME=EDIT
//EDIT      DD &EDIT,DCB=BLKSIZE=80
//*       FORT
//FORT    EXEC PGM=&FORTRAN,PARM='&COPTION',COND=(4,LT,EDIT)
//SYSPRINT  DD &CPRINT
//SYSPUNCH  DD DUMMY
//SYSUT1    DD UNIT=VIO,SPACE=(TRK,(3,3))
//SYSLIN    DD DSN=&LOADSET,DISP=(NEW,PASS),UNIT=SYSDA,
//            SPACE=(TRK,(3,3)),DCB=BLKSIZE=3200
//SYSIN     DD DSN=&TRANSFER,DISP=(OLD,DELETE)
//          DD DSN=&CLIB(&CMEM),DISP=SHR,LABEL=(,,,IN)
//*       LKED
//LKED    EXEC PGM=IEWLF128,COND=((4,LT,EDIT),(4,LT,FORT)),
//            PARM='&LOPTION,SIZE=(184K,60K)'
//SYSPRINT  DD &LPRINT,DCB=RECFM=UA
//SYSLIB    DD DSNAME=SYS2.FORTGH.LOADLIB,DISP=SHR
//          DD DSN=&LLIB,DISP=SHR
//          DD DSNAME=SYS2.FORTRAN.LOADLIB,DISP=SHR
//SYSUT1    DD UNIT=VIO,SPACE=(TRK,(57,19))
//SYSLMOD   DD DSN=&CFFSET(CFF),DISP=(NEW,PASS),UNIT=SYSDA,
//            SPACE=(CYL,(3,1,1),RLSE)
//SYSLIN    DD DSN=&LOADSET,DISP=(OLD,DELETE)
//          DD DSN=&CLIB(&LMEM),DISP=SHR
//*       GO
//CFF     EXEC PGM=CFF,
```

```
//                COND=((4,LT,EDIT),(4,LT,FORT),(4,LT,LKED))
//STEPLIB   DD DSN=&CFFSET,DISP=(OLD,PASS),UNIT=SYSDA
//FT05F001  DD DDNAME=INPUT
//FT06F001  DD DDNAME=LOG6
//FT02F001  DD DDNAME=LOG2
//FT03F001  DD DDNAME=LOG3
//FT04F001  DD DUMMY
//FT08F001  DD UNIT=VIO,DCB=(RECFM=VBS,LRECL=1612,BLKSIZE=1616),
//                SPACE=(TRK,(1,3))
//FT09F001  DD UNIT=VIO,DCB=(RECFM=VBS,LRECL=1612,BLKSIZE=1616),
//                SPACE=(TRK,(1,3))
//FT10F001  DD DDNAME=PARLIB
//FT11F001  DD UNIT=VIO,DCB=(RECFM=VBS,LRECL=588,BLKSIZE=592),
//                SPACE=(TRK,(1,3))
//FT12F001  DD DCB=(RECFM=VBS,LRECL=3092,BLKSIZE=3096),
//                SPACE=(TRK,(2,3)),UNIT=VIO
//FT13F001  DD UNIT=VIO,DCB=(RECFM=VBS,LRECL=556,BLKSIZE=560,
//                BUFNO=2),SPACE=(TRK,(1,1))
//FT14F001  DD UNIT=VIO,DCB=(RECFM=VBS,LRECL=X,BLKSIZE=13030,
//                BUFNO=2),SPACE=(TRK,(19,19),RLSE)
//FT15F001  DD UNIT=VIO,DCB=(RECFM=VBS,LRECL=X,BLKSIZE=13030,
//                BUFNO=2),SPACE=(TRK,(38,19),RLSE)
//FT16F001  DD UNIT=VIO,DCB=(RECFM=VBS,LRECL=4249,BLKSIZE=4253),
//                SPACE=(TRK,(2,3))
//FT17F001  DD DCB=(RECFM=VBS,LRECL=2055,BLKSIZE=2059),
//                SPACE=(TRK,(1,3)),UNIT=VIO
//FT18F001  DD UNIT=VIO,DCB=(RECFM=VBS,LRECL=1612,BLKSIZE=1616),
//                SPACE=(TRK,(10,3),RLSE)
//FT19F001  DD UNIT=VIO,DCB=(RECFM=VBS,LRECL=1612,BLKSIZE=12676,
//                BUFNO=2),SPACE=(TRK,(38,19),RLSE)
//FT20F001  DD DSN=&CORLIB(&COR),DISP=SHR,LABEL=(,,,IN)
//FT21F001  DD UNIT=VIO,DCB=(RECFM=VBS,LRECL=1612,BLKSIZE=1616),
//                SPACE=(TRK,(1,3))
//FT22F001  DD UNIT=VIO,DCB=(RECFM=VBS,LRECL=1612,BLKSIZE=12676,
//                BUFNO=2),SPACE=(TRK,(38,19),RLSE)
//FT23F001  DD UNIT=VIO,DCB=(RECFM=VBS,LRECL=540,BLKSIZE=544),
```

```
//              SPACE=(TRK,(1,3))
//FT24F001   DD UNIT=VIO,DCB=(RECFM=VBS,LRECL=1612,BLKSIZE=12676,
//              BUFNO=2),SPACE=(TRK,(38,19),RLSE)
//FT25F001   DD DSN=&INPUTS(&EXPIN),DISP=SHR,LABEL=(,,,IN)
//FT27F001   DD DSN=A108001.&ORT,DISP=(MOD,KEEP)
//FT37F001   DD DSN=A108001.&MST,DISP=(MOD,KEEP)
//LOG6       DD &LOG6,DCB=RECFM=UA
//LOG2       DD &LOG2,DCB=RECFM=UA
//LOG3       DD &LOG3,DCB=RECFM=UA
//PARLIB     DD DSN=&PARLIB(&PAM),DISP=SHR,LABEL=(,,,IN)
//INPUT      DD DSN=&INPUTS(DUAL),DISP=SHR,LABEL=(,,,IN)
//CFFLOG     DD DSN=A108001.CFFLOG,DISP=MOD
```

I have constructed a command list, which we use for small runs of
up to a few minutes CPU time. This became necessary when the IBM
3033 was so loaded with time sharing that even very small batch runs
might have a turn-around time of several hours. On the 3081 it is
not needed, but very convenient. The command list is interactive,
the program of course not. Only minimal output is taken to the
screen; most is directed to a fast printer.

The command list is reproduced below.

```
PROC 0 PARLIB(´A108001.PARAM´) CORLIB(´A108001.KOORD´) +
       EMEM(DUMMY) CMEM(DUMMY) FORTRAN(IEKAA00) +
       LMEM(NADIA0) INP(DUMMY) EXPIN(DUMMY) ORT(DUM) MST(DUM) +
       LOG2(N) GEM GEMT
CONTROL PROMPT MAIN END(SLUT)
ATTN EXIT
TIME
     WRITE
     WRITE                    ************************
     WRITE                    *    C F F    KALDT    *
     WRITE                    ************************
     WRITE
ATTRIB TRYK RECFM(F B A) LRECL(133) BLKSIZE(3990)
ALLOC REUSE F(UDPUT) DA(UDPUT) NEW CYL SPACE(1 1) USING(TRYK)
   IF &GEMT=GEMT THEN GOTO I
     WRITENR SKAL DER EDITERES I EN KILDETEKST I   S O U R C E ? :
     READ SVT
   IF &STR(&SVT)=END THEN GOTO S
   IF &STR(&SVT)=&STR(N) THEN GOTO C
     WRITENR PAA HVILKET MEDLEM?  :
     READ EMEM
     WRITENR HAR DU ALLEREDE EN EDIT-FIL?  :
     READ SV
   IF &STR(&SV)=&STR(J) THEN GOTO E
```

```
        WRITE HER STARTER EN EDIT FIL. TAST SELV VIDERE.
     QED EDIT NEW
     END SAVE
E:      WRITENR SKAL DER RETTES HERI ?  :
        READ SVE
     IF &STR(&SVE)=END THEN GOTO S
     IF &STR(&SVE)=&STR(N) THEN GOTO C
        WRITE RET! - AFSLUT MED KOMMANDOEN: SLUT
     QED EDIT
     TERMIN SLUT
     END SAVE
C:      WRITENR E D I T O R   ARBEJDER
ALLOC REUSE F(SYSPRINT) SYSOUT(A) HOLD
ALLOC REUSE F(SYSUT1) DA(´A108001.SOURCE(&EMEM)´) SHR
ATTRIB EDI RECFM(F B) LRECL(80) BLKSIZE(3120)
ALLOC REUSE F(SYSUT2) DA(TRANSFER) NEW TRA SPACE(1 3) USING(EDI)
ALLOC REUSE F(SCRATCH) NEW SPACE(57 19) TRA
     IF &STR(&SVT)=N THEN DO
ALLOC REUSE F(SYSIN) DA(DUM) SHR
     SLUT
     ELSE DO
ALLOC REUSE F(SYSIN) DA(EDIT) SHR
     SLUT
     CALL ´A108001.LINK(EDITOR)´ ´SS´
        WRITE  MED FEJLKODE &LASTCC
        WRITENR SKAL DER SUBSTITUERES FRA   S O U R C E ?  :
        READ SVC
     IF &STR(&SVC)=END THEN GOTO S
     IF &STR(&SVC)=&STR(N) THEN GOTO F
        WRITENR MED HVILKET MEDLEM ?  :
        READ CMEM
F:      WRITENR F O R T R A N   H   KALDES.
ALLOC REUSE F(SYSPRINT) SYSOUT(A) HOLD
ALLOC REUSE F(SYSUT1) NEW
ALLOC REUSE F(SYSLIN) DA(LOADSET) NEW SPACE(12 12) BLOCK(3120)
ALLOC REUSE F(SYSIN) DA(TRANSFER +
```

```
                      ´A108001.SOURCE(&CMEM)´) SHR
   CALL ´LNK2.FORTGH.LOAD(&FORTRAN)´ ´NOSOURCE,NOMAP´
      WRITE  FEJLKODE &LASTCC
I:    WRITENR SKRIV NAVN PAA INPUT  :
      READ INP
      WRITENR SKAL DER RETTES HERI ?  :
      READ SVI
   IF &STR(&SVI)=END THEN GOTO S
   IF &STR(&SVI)=&STR(N) THEN GOTO P
      WRITE RET! - AFSLUT MED KOMMANDOEN: SLUT
   QED ´A108001.INPUT(&INP)´
   TERMIN SLUT
   END SAVE
P:    WRITENR SKAL DER BRUGES PARAMETRE FRA &PARLIB ?  :
      READ SV
   IF &STR(&SV)=END THEN GOTO S
   IF &STR(&SV)=&STR(J) THEN GOTO PAR
      WRITE SKRIV NAVNET PAA PARAMETERBIBLIOTEKET  :
      READ PARLIB
PAR:  WRITENR HVILKET MEDLEM ?  :
      READ PAM
      WRITENR SKAL DER BRUGES KOORDINATER FRA &CORLIB ?  :
      READ SV
   IF &STR(&SV)=END THEN GOTO S
   IF &STR(&SV)=&STR(J) THEN GOTO K
      WRITE SKRIV NAVNET PAA KOORDINATBIBLIOTEKET  :
      READ CORLIB
K:    WRITENR HVILKET MEDLEM ?  :
      READ COR
      WRITENR SKAL DER OPTIMERES ?  :
      READ SVO
   IF &STR(&SVO)=END THEN GOTO S
   IF &STR(&SVO)=&STR(N) THEN GOTO L
      WRITE HVILKET MEDLEM I   I N P U T   SKAL BRUGES SOM EXPIN ? :
      READ EXPIN
      WRITENR SKAL DER RETTES HERI ?  :
```

```
       READ SV
   IF &STR(&SV)=END THEN GOTO S
   IF &STR(&SV)=&STR(N) THEN GOTO L
       WRITE RET! - AFSLUT MED KOMMANDOEN: SLUT
   QED ´A108001.INPUT(&EXPIN)´
   TERMIN SLUT
   END SAVE
L: IF &GEMT=GEMT THEN GOTO G
       WRITENR NU LINKER VI; VENT!
ALLOC REUSE F(SYSPRINT) SYSOUT(A) HOLD
ALLOC REUSE F(SYSLIB) DA(´SYS2.FORTGH.LOADLIB´ +
                         ´A108001.LINK´ +
                         ´SYS2.FORTRAN.LOADLIB´) SHR
ALLOC REUSE F(SYSUT1) NEW
ALLOC REUSE F(SYSLMOD) DA(CFFSET.LOAD(CFF)) NEW CYL SPACE(3 1) +
       DIR(1) RELEASE
ALLOC REUSE F(SYSLIN) DA(LOADSET +
                         ´A108001.SOURCE(&LMEM)´) SHR
   CALL ´SYS1.LINKLIB(IEWLF128)´ ´NOLIST,NOMAP´
       WRITE   - DET BLEV MED FEJLKODE &LASTCC
   GOTO GO
G: ALLOC REUSE DA(CFFSET.LOAD(CFF)) OLD
GO: ALLOC REUSE F(FT05F001) DA(´A108001.INPUT(&INP)´) SHR
   IF &STR(&SVO)=&STR(N) THEN DO
ALLOC REUSE F(FT06F001) DA(UDPUT) MOD
ALLOC REUSE F(FT03F001) DUMMY
   SLUT
   ELSE DO
ALLOC REUSE F(FT06F001) DUMMY
ALLOC REUSE F(FT03F001) DA(UDPUT) MOD
   SLUT
ALLOC REUSE F(FT04F001) DA(*)
   IF &STR(&LOG2)=&STR(J) THEN DO
ALLOC REUSE F(FT02F001) DA(UDPUT) MOD
   SLUT
   ELSE DO
```

```
ALLOC REUSE F(FT02F001) DUMMY
   SLUT.
ALLOC F(FT08F001) NEW
ALLOC F(FT09F001) NEW
ALLOC F(FT10F001) DA('&PARLIB(&PAM)') OLD
ALLOC F(FT11F001) NEW
ALLOC F(FT12F001) NEW
ALLOC F(FT13F001) NEW
ALLOC F(FT14F001) CYL SPACE(1 1)
ALLOC F(FT15F001) CYL SPACE(2 2)
ALLOC F(FT16F001) NEW
ALLOC F(FT17F001) NEW
ALLOC F(FT18F001) NEW
ALLOC F(FT19F001) CYL SPACE(2 1)
ALLOC F(FT20F001) DA('&CORLIB(&COR)') OLD
ALLOC F(FT21F001) NEW
ALLOC F(FT22F001) CYL SPACE(2 1)
ALLOC F(FT23F001) NEW
ALLOC F(FT24F001) CYL SPACE(2 1)
ALLOC F(FT25F001) DA('A108001.INPUT(&EXPIN)') SHR
ALLOC F(FT27F001) DA(&ORT) MOD
ALLOC F(FT37F001) DA(&MST) MOD
ALLOC F(CFFLOG) DA('A108001.CFFLOG') MOD
TIME
      WRITE NU REGNER VI.
   CALL CFFSET(CFF)
      WRITE   C F F   SLUT MED FEJLKODE &LASTCC
TIME
      WRITENR SKAL UDPUT TRYKKES ? :
      READ SVT
   IF &STR(&SVT)=N THEN GOTO U
   RENAME UDPUT UD
   SUBMIT JOB(UD)
U: IF &STR(&SVE)=J THEN P N EDIT
   IF &STR(&SVI)=J THEN P I &INP
S: ST A
```

```
    OUTSCAN *
    B
    -5/0
    IF &GEMT=GEMT THEN GOTO SL
ALLOC F(O) DELETE REUSE DA(TRANSFER)
ALLOC F(O) DELETE REUSE DA(LOADSET)
    WRITENR SKAL LADEMODUL SLETTES ?  :
    READ SVLM
    IF &STR(&SVLM)=&STR(J) THEN ALLOC F(O) DELETE REUSE +
    DA(CFFSET.LOAD)
SL:   WRITENR SKAL UDPUT SLETTES ?  :
    READ SVUT
    IF &STR(&SVUT)=&STR(J) THEN ALLOC F(O) DELETE REUSE DA(UDPUT)
    WRITENR SKAL UD SLETTES ?  :
    READ SVUD
    IF &STR(&SVUD)=&STR(J) THEN ALLOC F(O) DELETE REUSE DA(UD)
    FREE ALL
ACCOUNT SESSION
```

Appendix 6   The future of CFF

Will you still need me
will you still feed me
when I'm sixty-four

The Beatles: Sgt. Pepper

When concluding a book which to some extent is a review of one's own work, it is tempting to prophesize a bit, or anyway to express good wishes for oneself and associates. Let me therefore sketch future developments in our methods and programs as I hope and intend to see them.

The existing program will slowly be perfected and expanded. Calculation of NMR chemical shifts will be added, IR circular dichroism is already there, optimisation on thermodynamic differences may be introduced.

The program will be used to optimise simple, few-parameter functions on more classes of compounds; in addition to those mentioned before: acids, esters, amides, amino acids, ethenes, aromatics, halogen compounds. Based on these results, it will be used for prediction. In the long run, as the new program (see below) develops, it will be used more exclusively on coordination complexes, in solution and in crystals.

A new generation of the consistent force field program is being born. It is written in pascal, allows for - in principle - any number of atoms, and can distinguish and handle twenty-odd types of atoms in the same run. It can read and interpret the same formula types as the present FORTRAN program, but will be able also to understand super-formulae, the units of which are saccharides, peptides and the genetic alphabet. The program will employ a powerful minimiser and be linked to graphical systems. It will not encompass optimisation but will use the results of the older generation in the modelling of biopolymers. It will include a self-assembly procedure which can build up biological structures. It will not be used in secondary and tertiary polypeptide structure research: so many are already proficient in this.

The new program is intended to be easily portable, so that it can be implemented on any computer, also on the new micros with a word length of 32 bits. It will not be available for quite a few years, which will - or should - sadden drug designers.

Literature references

(1)   A.A. Abdurahmanov, R.A. Rahimova and L.M. Imanov
        Phys. Lett. 32A (1970) 123-124

(2)   A.A. Abdurahmanov, R.A. Rahimova, E.I. Veliyulin and L.M. Ima-
        nov
        Dokl. Akad. Nauk Azerbad. SSR 32 (1976) 14-17

(3)   N.L. Allinger
        Adv. Phys. Org. Chem. 13 (1976) 1-84

(4)   A. Almenningen, O. Bastiansen and P.N. Skancke
        Acta Chem. Scand. 15 (1961) 711-712

(5)   A. Almenningen, H.M. Seip and T. Willadsen
        Acta Chem. Scand. 23 (1969) 2748-2754

(6)   C. Altona and D.H. Faber
        Top. Curr. Chem. 45 (1974) 1-38

(7)   J.E. Anderson
        Fortschr. d. Chem. Forsch. 45 (1974) 139-167

(8)   J.D. Andose and K. Mislow
        J. Am. Chem. Soc. 96 (1974) 2168-2176

(9)   F. Arène, A. Neuman and F. Longchambon
        C. R. Acad. Sc. Paris 288C (1979) 331-334

(10)   E.E. Astrup
        Acta Chem. Scand. 27 (1973) 1345-1350

(11)   E.E. Astrup
        Acta Chem. Scand. 27 (1973) 3271-3276

(12)   E.E. Astrup and A.M. Aomar
        Acta Chem. Scand. A29 (1975) 794-798

(13)   J.M.A. Baas, B. van de Graaf, A. van Veen and B.M. Wepster
        Tetrahedron Lett. (1978) 819-820

(14)   J.M.A. Baas, B. van de Graaf, D. Tavernier and P. Vanhee
        J. Am. Chem. Soc. 103 (1981) 5014-5021

(15)   L.S. Bartell and H.B. Bürgi
        J. Am. Chem. Soc. 94 (1972) 5239-5246

(16)   L.S. Bartell and T.L. Boates
        J. Mol. Struct. 32 (1976) 379-392

(17)   L.S. Bartell and W.F. Bradford
        J. Mol. Struct. 37 (1977) 113-126

(18)   O. Bastiansen, L. Fernholt, H.M. Seip, H. Kambara and K.
        Kuchitsu
        J. Mol. Struct. 18 (1973) 163-168

(19)   Bauder
        J. Phys. Chem. Ref. Data 8 (1979) 583-618

(20) B. Beagley, D.P. Brown and J.J. Monaghan
      J. Mol. Struct. 4 (1969) 233-244

(21) B. Beagley
      Mol. Struct. by Diffraction Meth. 6 (1978) 63-92

(22) R.W. Berg and I. Søtofte
      Acta Chem. Scand. A30 (1976) 843-844

(23) T.L. Bluhm, Y. Deslandes, R.H. Marchessault, S. Perez and M.
      Rinaudo
      Carbohydr. Res. 100 (1982) 117-130

(24) K. Bock and M. Vignon
      Nouv. J. Chim. 6 (1982) 301-304

(25) K. Bock
      Pure Appl. Chem. 55 (1983) 605-622

(26) J.C.A. Boeyens, R.D. Hancock and G.J. McDougall
      S. Afr. J. Chem. 32 (1979) 23-26

(27) G. Bolis and E. Clementi
      J. Am. Chem. Soc. 99 (1977) 5550-5557

(28) M.J. Bovill, D.J. Chadwick, I.O. Sutherland and D. Watkin
      J. Chem. Soc. Perkin Trans. 2 (1980) 1529-1543

(29) R.H. Boyd
      J. Chem. Phys. 49 (1968) 2574-2583

(30) R.H. Boyd,S.M. Breitling and M. Mansfield
      AIChE J. 19 (1973) 1016-1024

(31) W.F. Bradford, S. Fitzwater and L.S. Bartell
      J. Mol. Struct. 38 (1977) 185-194

(32) J.L. Bredas, M. Dufey, J.G. Fripiat and J.M. André
      Mol. Phys. 49 (1983) 1451-1460

(33) H.E. Breed, G. Gundersen and R. Seip
      Acta Chem. Scand. A33 (1979) 225-233

(34) F. Brisse, R.H. Marchessault, S. Perez and P. Zugenmaier
      J. Am. Chem. Soc. 104 (1982) 7470-7476

(35) B.R. Brooks, R.E. Bruccoleri, B.D. Olafson, D.J. States, S.
      Swaminatan and M. Karplus
      J. Comput. Chem. 4 (1983) 187-217

(36) J.S. Brown
      Proc. Phys. Soc. 89 (1966) 987-992

(37) G.R. Brubaker and R.A. Euler
      Inorg. Chem. 11 (1972) 2357-2361

(38) G.R. Brubaker and J.G. Massura
      J. Coord. Chem. 3 (1974) 251-251

(39) D.A. Buckingham, I.E. Maxwell, A.M. Sargeson and M.R. Snow
      J. Am. Chem. Soc. 92 (1970) 3617-3626

(40)   D.A. Buckingham and A.M. Sargeson
       Topics in Stereochem. 6 (1971) 219-277

(41)   P. Bugnon and C.W. Schlaepfer
       Inorg. Chim. Acta 40 (1980) 108-109

(42)   A.W. Burgess, L.L. Shipman and H.A. Scheraga
       Proc. Nat. Acad. Sci. 72 (1975) 854-858

(43)   A.W. Burgess, L.L. Shipman, R.A. Nemenoff and H.A. Scheraga
       J. Am. Chem. Soc. 98 (1976) 23-29

(44)   H.B. Bürgi and L.S. Bartell
       J. Am. Chem. Soc. 94 (1972) 5236-5238

(45)   U. Burkert and N.L. Allinger
       Molecular Mechanics, ACS Monograph 177 (1982)

(46)   V. Busetti, M. Mammi and G. Carazzolo
       Z. Krist. 119 (1963) 310-318

(47)   V. Busetti, A.del Pra and M. Mammi
       Acta Crystallogr. B25 (1969) 1191-1194

(48)   A.H. Clark and T.G. Hewitt
       J. Mol. Struct. 9 (1971) 33-47

(49)   E. Clementi
       Determination of Liquid Water Structure, Coordination Numbers
       for Ions and Solvation for Biological Molecules. Lecture Notes
       in Chemistry, Vol. 2, Springer - Verlag (1976)

(50)   E. Clementi, F. Cavallone and R. Scordamaglia
       J. Am. Chem. Soc. 99 (1977) 5531-5545

(51)   J.L. De Coen, G. Elefante, A.M. Liquori and A. Damiani
       Nature 216 (1967) 910-913

(52)   E.J. Corey and J.C. Bailar, Jr.
       J. Am. Chem. Soc. 81 (1959) 2620-2629

(53)   G. Corongiu and E. Clementi
       Gazz. Chim. Ital. 108 (1978) 273-306

(54)   C.A. Coulson and U. Danielsson
       Arkiv for Fysik 8 (1954) 239-244, 245-255

(55)   P. Dais and A.S. Perlin
       Carbohydr. Res. 107 (1982) 263-269

(56)   J. Dale
       Stereochemistry and Conformational Analysis, Verlag Chemie,
       New York & Weinheim (1978)

(57)   M. Davis and O. Hassel
       Acta Chem. Scand. 17 (1963) 1181

(58)   L.J. DeHayes
       Thesis, Univ. Microfilms 72-20,952 (1972)

(59)   L.J. DeHayes and D.H. Busch
       Inorg. Chem. 12 (1973) 1505-1513

(60) S.K. Doun and L.S. Bartell
J. Mol. Struct. 63 (1980) 249-258

(61) A. Dunand and R. Gerdil
Acta Crystallogr. B31 (1975) 370-374

(62) L.G. Dunfield, A.W. Burgess and H.A. Scheraga
J. Phys. Chem. 82 (1978) 2609-2616

(63) J.D. Dunitz and H. Eser
Helv. Chim. Acta 50 (1967) 1565-1572

(64) J.D. Dunitz
Perspectives in Struct. Chem (J.D. Dunitz and J.A. Ibers, Eds.), Vol. 2 Wiley, NY (1968) p. 1-70

(65) J.D. Dunitz and H.-B. Bürgi
MTP Int. Rev. Sci. 11 (1976) 81-120

(66) J.R. Durig, Y.S. Li and C.C. Tong
J. Mol. Struct. 18 (1973) 269-275

(67) J.R. Durig and D.A.C. Compton
J. Phys. Chem. 83 (1979) 265-268

(68) M. Dwyer, R.J. Geue and M.R. Snow
Inorg. Chem. 12 (1973) 2057-2061

(69) E.L. Eliel, N.L. Allinger, S.J. Angyal and G.A. Morrison
Conformational Analysis, Wiley, NY (1965) p. 446-460

(70) E.M. Engler, J.D. Andose and P. von R. Schleyer
J. Am. Chem. Soc. 95 (1973) 8005-8025

(71) O. Ermer and S. Lifson
J. Am. Chem. Soc. 95 (1973) 4121-4132

(72) O. Ermer
Struct. Bonding 27 (1976) 161-211

(73) O. Ermer
Aspekte von Kraftfeldrechnungen, Wolfgang Bauer Verlag, München (1981)

(74) D.H. Faber and C. Altona
Computers & Chem. 1 (1977) 203-213

(75) C.K. Fair and E.O. Schlemper
Acta Crystallogr. B33 (1977) 1337-1341

(76) S. Fitzwater and L.S. Bartell
J. Am. Chem. Soc. 98 (1976) 5107-5115

(77) S. Fraga
J. Comput. Chem. 3 (1982) 329-334

(78) M.F. Gargallo, R.E. Tapscott and E.N. Duesler
Inorg. Chem. 23 (1984) 918-922

(79) N. Gavrushenko, H.L. Carrell, W.C. Stallings and J.P. Glusker
Acta Crystallogr. B33 (1977) 3936-3939

(80) B.R. Gelin and M. Karplus
      J. Am. Chem. Soc. 97 (1975) 6996-7006

(81) R.J. Geue and M.R. Snow
      J. Chem. Soc. A (1971) 2981-2987

(82) G. Giunchi and L. Barino
      Gazz. Chim. Ital. 110 (1980) 395-401

(83) C.P.J. Glaudemans, P. Kovac and Kj. Rasmussen
      Biochemistry in print

(84) J.R. Gollogly and C.J. Hawkins
      Inorg. Chem. 8 (1969) 1168-1173

(85) J.R. Gollogly and C.J. Hawkins
      Inorg. Chem. 9 (1970) 576-582

(86) J.R. Gollogly, C.J. Hawkins and J.K. Beattie
      Inorg. Chem. 10 (1971) 317-323

(87) J.R. Gollogly and C.J. Hawkins
      Inorg. Chem. 11 (1972) 156-161

(88) B. van de Graaf, J.M.A. Baas and A. van Veen
      Recl. Trav. Chim. Pays-Bas 99 (1980) 175-178

(89) B. van de Graaf, J.M.A. Baas and M.A. Widya
      Recl. Trav. Chim. Pays-Bas 100 (1981) 59-61

(90) G. Gundersen
      Personal communication (1980)

(91) W.F. van Gunsteren and M. Karplus
      Nature 293 (1981) 677-678

(92) A.T. Hagler and S. Lifson
      Acta Cryst. B30 (1974) 1336-1341

(93) A.T. Hagler, E. Huler and S. Lifson
      J. Am. Chem. Soc. 96 (1974) 5319-5327

(94) A.T. Hagler and S. Lifson
      J. Am. Chem. Soc. 96 (1974) 5327-5335

(95) A.T. Hagler, L. Leiserowitz and M. Tuval
      J. Am. Chem. Soc. 98 (1976) 4600-4612

(96) A.T. Hagler, P.S. Stern, S. Lifson and S. Ariel
      J. Am. Chem. Soc. 101 (1979) 813-819

(97) A.T. Hagler, S. Lifson and P. Dauber
      J. Am. Chem. Soc. 101 (1979) 5122-5130

(98) A.T. Hagler, P. Dauber and S. Lifson
      J. Am. Chem. Soc. 101 (1979) 5131-5141

(99) N.C.P. Hald and Kj. Rasmussen
      Acta Chem. Scand. A32 (1978) 753-756

(100) N.C.P. Hald and Kj. Rasmussen
      Acta Chem. Scand. A32 (1978) 879-886

(101) T.W. Hambley, C.J. Hawkins, J.A. Palmer and M.R. Snow
      Aust. J. Chem. 34 (1981) 45-56

(102) R.D. Hancock, G.J. McDougall and F. Marsicano
      Inorg. Chem. 18 (1979) 2847-2852

(103) R.D. Hancock, G.J. McDougall
      J. Am. Chem. Soc. 102 (1980) 6551-6553

(104) I. Harada, M. Takeuchi, H. Sakakibara, H. Matsuura and T. Shi-
      manouchi
      Bull. Chem. Soc. Japan 50 (1977) 102-110

(105) M. Hayashi and K. Kuwada
      Bull. Chem. Soc. Japan 47 (1974) 3006-3009

(106) M. Hayashi and K. Kuwada
      J. Mol. Struct. 28 (1975) 147-161

(107) J.B. Hendrickson
      J. Amer. Chem. Soc. 83 (1961) 4537-4547

(108) R.L. Hilderbrandt and J.D. Wieser
      J. Mol. Struct. 15 (1973) 27-36

(109) C.J. Hilleary, T.F. Them and R.E. Tapscott
      Inorg. Chem. 19 (1980) 102-107

(110) S. Hirokawa, M. Masakuni, M. Seki and N. Teruo
      Mem. Def. Acad. (Math., Phys., Chem. Eng.) 8 (1968) 485-498

(111) Z.I. Hodes, G. Nemethy and H.A. Scheraga
      Biopolymers 18 (1979) 1561-1610

(112) A.J. Hopfinger
      Conformational Properties of Macromolecules, Academic Press
      (1973)

(113) E. Huler and A. Warshel
      Acta Cryst. B30 (1974) 1822-1826

(114) M.G. Hutchings, J.G. Nourse and K. Mislow
      Tetrahedron 30 (1974) 1535-1549

(115) M.G. Hutchings, J.D. Andose and K. Mislow
      J. Am. Chem. Soc. 97 (1975) 4553-4561

(116) M.G. Hutchings, J.D. Andose and K. Mislow
      J. Am. Chem. Soc. 97 (1975) 4562-4570

(117) T. Iijima
      Bull. Chem. Soc. Japan 45 (1972) 1291-1294

(118) T. Iijima
      Bull. Chem. Soc. Japan 46 (1973) 2311-2314

(119) S.A. Islam and S. Neidle
      Acta Cryst. B39 (1983) 114-119

(120) K. Ito
      J. Am. Chem. Soc. 75 (1953) 2430-2435

(121) J. Jacob, H.B. Thompson and L.S. Bartell
J. Chem. Phys. 47 (1967) 3736-3753

(122) C. Jaime and E. Osawa
Tetrahedron 39 (1983) 2769-2778

(123) S. Jamet-Delcroix
Acta Crystallogr. B29 (1973) 977-980

(124) O. Jardetzky
Biochim. Biophys. Acta 621 (1980) 227-232

(125) P.G. Jones, G.M. Sheldrick, A. J. Kirby and R. Glenn
Z. Krist. 161 (1982) 237-243, 245-251

(126) N.V. Joshi and V.S.R. Rao
Biopolymers 18 (1979) 2993-3004

(127) C.O. Kadzhar, A.A. Ababsov, A.B. Askerov and L.M. Imanov
Izv. Akad. Nauk Azerb. SSR, Ser. Fiziko-tekhn. Mat. Nauk
(1973) 80-84

(128) R.K. Kakar and P.J. Seibt
J. Chem. Phys. 57 (1972) 4060-4061

(129) H. Kato, J. Nakagawa and M. Hayashi
J. Mol. Spectr. 80 (1980) 272-278

(130) S.K. Katti, T.P. Seshadri and M.A. Viswamitra
Acta Cryst. B38 (1982) 1136-1140

(131) R. Kewley
J. Mol. Spectr. 89 (1981) 548-555

(132) K. Kildeby, S. Melberg and Kj. Rasmussen
Acta Chem. Scand. A31 (1977) 1-13

(133) K. Kimura and M. Kubo
J. Chem. Phys. 30 (1959) 151-158

(134) A.I. Kitaigorodsky
Tetrahedron 9 (1960) 183-193

(135) A.I. Kitaigorodsky
Tetrahedron 14 (1961) 230-238

(136) A.I. Kitaigorodsky
Molecular Crystals and Molecules, Academic Press, New York
(1973)

(137) A.I.Kitaigorodsky
Mixed Crystals, Springer-Verlag, Heidelberg (1984)

(138) S. Kondo and E. Hirota
J. Mol. Spectr. 34 (1970) 97-107

(139) L.M.J. Kroon-Batenburg and J.A. Kanters
Acta Cryst. B39 (1983) 749-754

(140) K. Kuchitsu
Phys. Chem. Ser. 1, Vol. 2, MTP Int. Rev. Sci., Butterworths
(1972), p. 203-239

(141) K. Kuchitsu and S.J. Cyvin
Molecular Structures and Vibrations (S.J.Cyvin, Ed.), Else-
vier, Amsterdam (1972) chap. 12, p. 183-211

(142) K. Kuchitsu
Critical Evaluation of Chem. and Phys. Struct. Inform. (D.R.
Lide and M.A. Paul, Eds.), Nat. Acad. Sci., Washington (1974),
p. 132-139

(143) T. Laier and E. Larsen
Acta Chem. Scand. A33 (1979) 257-264

(144) Landolt-Bornstein
6. Aufl. 2. Band 4. Teil, Springer, Berlin (1961) p. 461

(145) J. Larsen
Unpublished work (1979)

(146) R.M. Lees and J.G. Baker
J. Chem. Phys. 48 (1968) 5299-5318

(147) R.U. Lemieux and K. Bock
Arch. Biochem. Biophys. 221 (1983) 125-134

(148) F. Leung, H.D. Chanzy, S. Peres and R.H. Marchessault
Can J. Chem. 54 (1976) 1365-1371

(149) R.M. Lees, F.J. Lovas, W.H. Kirchhoff and D.R.Johnson
J. Phys. Chem. Ref. Data 2 (1973) 205-214

(150) D.R. Lide, Jr.
J. Chem. Phys. 33 (1960) 1514-1518

(151) S. Lifson and A. Warshel
J. Chem. Phys. 49 (1968) 5116-5129

(152) S. Lifson and M. Levitt
Computers & Chem. 3 (1979) 49-50

(153) S. Lifson, A.T. Hagler and P. Dauber
J. Am. Chem. Soc. 101 (1979) 5111-5121

(154) S. Lifson and P.S. Stern
J. Chem. Phys. 77 (1982) 4542-4550

(155) S. Lifson
Supramolecular Structure and Function (G. Pifat and J.N.
Herak, Eds.), Plenum Press, New York (1983), p. 1-44

(156) A.M. Liquori
J. Polymer Sci. Part C 12 (1966) 209-234

(157) A.M. Liquori, A. Damiani and G. Elefante
J. Mol. Biol. 33 (1968) 439-444

(158) M. Magini, G. Paschina and G. Piccaluga
J. Chem. Phys. 77 (1982) 2051-2056

(159) F.T. Marchese, P.K. Mehrota and D.L. Beveridge
J. Phys. Chem. 86 (1982) 2592-2601

(160) K.-M. Marstokk and H. Møllendal
J. Mol. Struct. 49 (1978) 221-237

(161) J.-P. Mathieu
Ann. de Phys. 19 (1944) 335-354

(162) H. Mathisen, N. Norman and B.F. Pedersen
Acta Chem. Scand. 21 (1967) 127-135

(163) O. Matsuoka, C. Tosi and E. Clementi
Biopolymers 17 (1978) 33-49

(164) E. Maverick, P. Seiler, W.B. Schweizer and J.D. Dunitz
Acta Cryst. B36 (1980) 615-620

(165) J.P. McCullogh, R.E. Pennington, J.C. Smith, I.A. Hossenlopp
and G. Waddington
J. Am. Chem. Soc. 81 (1959) 5880-5883

(166) G.J.McDougall, R.D. Hancock and J.C.A. Boeyens
J. Chem. Soc. Dalton Trans. (1978) 1438-1444

(167) S. Melberg and Kj. Rasmussen
Acta Chem. Scand. A32 (1978) 187-188

(168) S. Melberg and Kj. Rasmussen
Carbohydr. Res. 69 (1979) 27-38

(169) S. Melberg and Kj. Rasmussen
Carbohydr. Res. 71 (1979) 25-34

(170) S. Melberg, Kj. Rasmussen, R. Scordamaglia and C. Tosi
Carbohydr. Res. 76 (1979) 23-37

(171) S. Melberg and Kj. Rasmussen
J. Mol. Struct. 57 (1979) 215-239

(172) S. Melberg and Kj. Rasmussen
Carbohydr. Res 78 (1980) 215-224

(173) F.C. Mijlhoff, H.J. Geise and E.J.M. van Schaick
J. Mol. Struct.20 (1973) 393-401

(174) K. Mirsky
Acta Cryst. A32 (1976) 199-207

(175) H. Møllendal
personal communication (1979)

(176) F.A. Momany, L.M. Carruthers, R.F. McGuire and H.A. Scheraga
J. Phys. Chem. 78 (1974) 1595-1620

(177) F.A. Momany, R.F. McGuire, A.W. Burgess and H.A. Scheraga
J. Phys. Chem. 79 (1975) 2361-2381

(178) D.G. Montague, I.P. Gibson and J.C. Dore
Mol. Phys. 44 (1981) 1355-1367

(179) D.G. Montague, I.P. Gibson and J.C. Dore
Mol. Phys. 47 (1982) 1405-1416

(180) J.J. Moré
Lect. Notes in Math. 630 (1978) 105-116

(181) G. Nemethy and H.A. Scheraga
Quart. Rev. Biophys. 10 (1977) 239-352

(182) G. Nemethy, W.J. Peer and H.A. Scheraga
      Ann. Rev. Biophys. Bioeng. 10 (1981) 459-497

(183) G. Nemethy, M.S. Pottle and H.A. Scheraga
      J. Phys. Chem. 87 (1983) 1883-1887

(184) G.J.H. van Nes and A. Vos
      Acta Cryst. B34 (1978) 1947-1956

(185) B.R.A. Nijboer and F.W. de Wette
      Physica 23 (1957) 309-321

(186) S.R. Niketic and F. Woldbye
      Acta Chem. Scand. 27 (1973) 621-642

(187) S.R. Niketic and F. Woldbye
      Acta Chem. Scand. 27 (1973) 3811-3816

(188) S.R. Niketic and F. Woldbye
      Acta Chem. Scand. A28 (1974) 248

(189) S.R. Niketic, Kj. Rasmussen, F. Woldbye and S. Lifson
      Acta Chem. Scand. A30 (1976) 485-497

(190) S.R. Niketic and Kj. Rasmussen
      The Consistent Force Field: A Documentation. Lecture Notes in
      Chemistry, Vol. 3, Springer - Verlag (1977)

(191) S.R. Niketic and Kj. Rasmussen
      Acta Chem. Scand. A32 (1978) 391-400

(192) S.R. Niketic and Kj. Rasmussen
      Acta Chem. Scand. A35 (1981) 213-218

(193) S.R. Niketic and Kj. Rasmussen
      Acta Chem. Scand. A35 (1981) 623-633

(194) N. Norman and H. Mathisen
      Acta Chem. Scand. 15 (1961) 1755-1760

(195) J.M. O'Gorman, Jr., W. Shand and V. Schomaker
      J. Am. Chem. Soc. 72 (1950) 4222-4228

(196) T. Oka, K. Tsuchiya, s. Iwata and Y. Morino
      Bull. Chem. Soc. Japan 37 (1964) 4-7

(197) E. Osawa
      Personal communication (1979)

(198) E. Osawa, J.B. Collins and P.v.R. Schleyer
      Tetrahedron 33 (1977) 2667-2675

(199) K. Oyanagi and K. Kuchitsu
      Bull. Chem. Soc. Japan 51 (1978) 2237-2242

(200) K. Oyanagi and K. Kuchitsu
      personal communication (1983)

(201) F. Pavelčik and J. Majer
      Coll. Czech. Chem. Commun. 43 (1978) 1450-1459

(202) F. Pavelčik and J. Majer
      Coll. Czech. Chem. Commun. 47 (1982) 465-475

(203) M. Peereboom. B. van de Graaf and J.M.A. Baas
Recl. Trav. Chim. Phys-Bas 101 (1982) 336-338

(204) M.R. Peterson and I.G. Csizmadia
J. Am. Chem. Soc. 100 (1978) 6911-6916

(205) T. Philip, R.L. Cook, T.B. Malloy, Jr., N.L. Allinger, S. Chang and Y. Yuh
J. Am. Chem. Soc. 103 (1981) 2151-2156

(206) H.M. Pickett and H.L. Strauss
J. Am. Chem. Soc. 92 (1970) 7281-7290

(207) L.-O. Pietilä and Kj. Rasmussen
J. Comput. Chem. 5 (1984) 252-260

(208) D. Poland and H.A. Scheraga
Biochemistry 6 (1967) 3791-3800

(209) G.L. Pollack
Rev. Mod. Phys. 36 (1964) 748-791

(210) R. Poupko and Z. Luz
J. Chem. Phys. 75 (1981) 1675-1681

(211) G. Ranghino and E. Clementi
Gazz. Chim. Ital. 108 (1978) 157-170

(212) V.S.R. Rao, P.R. Sundararajan, C. Ramakrishnan and G.N. Ramachandran
Conformation of Biopolymers (G.N. Ramachandran, Ed.), Vol. 2, Academic Press, London (1967), p. 721-737

(213) Kj. Rasmussen
Conformations and vibrational spectra of tris(diamine) chelate complexes, Chemistry Department A (1970). Available from Technical Libary of Denmark, DK-2800 Lyngby

(214) Kj. Rasmussen and J. Larsen
Comment. Phys. - Math. Soc. Sci. Fenn. 48 (1978) 102-113

(215) Kj. Rasmussen
Molecular Structure and Dynamics, (M. Balaban, Ed.), Balaban, Jerusalem (1980) 171-210

(216) Kj. Rasmussen and F. Woldbye
Coordination Chemistry - 20 (D. Banerjea, Ed.), Pergamon Press, Oxford (1980), p. 219-227

(217) Kj. Rasmussen
J. Mol. Struct. 72 (1981) 171-176

(218) Kj. Rasmussen
Acta Chem. Scand. A36 (1982) 323-327

(219) Kj. Rasmussen
J. Mol. Struct. 97 (1983) 53-56

(220) Kj. Rasmussen and C. Tosi
Acta Chem. Scand. A37 (1983) 79-91

(221) Kj. Rasmussen and C. Tosi
      J. Mol. Struct. in print

(222) Kj. Rasmussen and B. Rasmussen
      manuscript

(223) Kj. Rasmussen
      manuscript

(224) Kj. Rasmussen
      manuscript

(225) K.N. Raymond, P.W.R. Corfield and J.A. Ibers
      Inorg. Chem. 7 (1968) 1362-1372

(226) K.N. Raymond, J.A. Ibers
      Inorg. Chem. 7 (1968) 2333-2338

(227) E. Th. Rietschel, U. Schade, M. Jensen, H.-W. Wollenweber, O.
      Lüderitz and S.G. Greisman
      Scand. J. Infect. Dis. Suppl. 31 (1982) 8-21

(228) D.C. Rohrer, A. Sarko, T.L. Bluhm and Y.N. Lee
      Acta Cryst. B36 (1980) 650-654

(229) A. Sabatini and S. Califano
      Spectrochim. Acta 16 (1960) 677-688

(230) Y. Saito
      Pure Appl. Chem. 17 (1968) 21-36

(231) Y. Saito
      Topics in Stereochem. 10 (1978) 95-174

(232) Y. Saito
      Inorganic Molecular Dissymmetry, Springer-Verlag, Berlin-Hei-
      delberg-New York (1979)

(233) Y. Sasada, M. Takano and T. Satoh
      J. Mol. Spectr. 38 (1971) 33-42

(234) J.A. Schellman and S. Lifson
      Biopolymers 12 (1973) 315-327

(235) H.A. Scheraga, R.A. Scott, G. Vanderkooi, S.J. Leach, K.D.
      Gibson and T. Ooi
      Conformation of Biopolymers (G.N. Ramachandran, Ed.), Vol. 1,
      Academic Pres, London (1967), p. 43-60

(236) H.A. Scheraga
      Adv. Phys. Org. Chem. 6 (1968) 103-184

(237) H.A. Scheraga
      Peptides - Proc. 5th Am. Peptide Symp. (M. Goodman and J. Mei-
      enhofer, Eds.) (1977) p. 246-256

(238) W.B. Schweizer and J.D. Dunitz
      Helv. Chem. Acta 65 (1982) 1547-1554

(239) R. Scordamaglia, F. Cavallone and E. Clementi
      J. Am. Chem. Soc. 99 (1977) 5545-5550

(240) R.A. Scott and H.A. Scheraga
J. Chem. Phys. 42 (1965) 2209-2215

(241) D.W. Scott
Chemical Thermodynamic Properties of Hydrocarbons, U. S.
Bureau of Mines Bull. 666 (1974)

(242) B. Sheldrick and D. Akrigg
Acta Cryst. C39 (1983) 315-316

(243) T. Shimanouchi
NSRDS-NBS 39 (1972)

(244) L.L. Shipman, A.W. Burgess and H.A. Scheraga
Proc. Nat. Acad. Sci. 72 (1975) 543-547

(245) L.L. Shipman, A.W. Burgess and H.A. Scheraga
J. Phys. Chem. 80 (1976) 52-54

(246) J. Snir, R.A. Nemenoff and H.A. Scheraga
J. Phys. Chem. 82 (1978) 2497-2530 (5 papers)

(247) M.R. Snow, D.A. Buckingham, P.A. Marzilli and A.M. Sargeson
Chem. Communic. (1969) 891-892

(248) M.R. Snow
J. Am. Chem. Soc. 92 (1970) 3610-3617

(249) M.R. Snow
J. Chem. Soc. Dalton Trans (1972) 1627-1634

(250) T.L. Starr and D.E. Williams
Acta Cryst. A33 (1977) 771-776

(251) S. Stessin and C. Clement
C. R. Acad. Sci. 276C (1973) 261-264

(252) A.J. Stuper, T.M. Dyott and G.S. Zander
Computer-assisted Drug Design (E.C. Olsen and R.E. Christof-
fersen, Eds.) ACS Symp. Ser. 112 (1979) p. 383-414

(253) T. Sundius
Commentat. Phys. - Math. Soc. Sci. Fenn. 47 (1977) 1-66

(254) T. Sundius and Kj. Rasmussen
Commentat. Phys. - Math. Soc. Sci. Fenn. 47 (1977) 91-98

(255) T. Sundius and Kj. Rasmussen
J. Mol. Struct. 65 (1980) 215-218

(256) K. Suzuki and K. Iguchi
J. Chem. Phys. 77 (1982) 4594-4603

(257) M. Takano, Y. Sasada and T. Satoh
J. Mol. Spectr. 26 (1968) 157-162

(258) H.-Y. Ting, W. H. Watson and H.C. Kelly
Inorg. Chem. 11 (1972) 374-377

(259) C. Tosi, E. Clementi and O. Matsuoka
Biopolymers 17 (1978) 51-66

(260) C. Tosi and E. Pescatori
      Gazz. Chim. Ital. 108 (1978) 365-372

(261) C. Tosi and Kj. Rasmussen
      Biopolymers 20 (1981) 1059-1088

(262) I. Tvaroska
      Biopolymers 21 (1982) 1887-1897

(263) P. Vanhee, B. van de Graaf, D. Tavernier and J.M.A. Baas
      J. Org. Chem. 48 (1983) 648-652

(264) A. Warshel and S. Lifson
      Chem. Phys. Lett. 4 (1969) 255-256

(265) A. Warshel, M. Levitt and S. Lifson
      J. Mol. Spectr. 33 (1970) 84-99

(266) A. Warshel and Lifson
      J. Chem. Phys. 53 (1970) 582-594

(267) A. Warshel
      J. Chem. Phys. 55 (1971) 3327-3335

(268) A. Warshel and M. Karplus
      J. Am. Chem. Soc. 94 (1972) 5613-5656

(269) A. Warshel
      Israel J. Chem. 11 (1973) 709-717

(270) A. Warshel and M. Karplus
      J. Am. Chem. Soc. 96 (1974) 5677-5689

(271) A. Warshel
      Modern Theor. Chem. 7 (1977) 133-172

(272) T. Wasiutynski, A. van der Avoird and R.M. Berns
      J. Chem. Phys. 69 (1978) 5288-5300

(273) P.K. Weiner and P.A. Kollman
      J. Comput. Chem. 2 (1981) 287-303

(274) S. Weiss and G.E. Leroi
      J. Chem. Phys. 48 (1968) 962-967

(275) R.C. West and M.J. Astle (Eds.)
      CRC Handbook of Chemistry and Physics, 62nd edition, CRC
      Press, Boka Raton, Florida (1981)

(276) F.H. Westheimer
      Steric Effects in Org. Chem. (M.S. Newman, Ed.), Wiley, NY
      (1956) p. 523-555

(277) D.N.J. White
      Mol. Struct. by Diffraction Meth. 6 (1978) 38-62

(278) K.B. Wiberg
      J. Am. Chem. Soc. 87 (1965) 1070-1078

(279) J.E. Williams, P.J. Stang and P.v.R. Schleyer
      Ann. Rev. Phys. Chem. 19 (1968) 531-558

(280) D.E. Williams
      Acta Cryst. A27 (1971) 452-455

(281) D.E. Williams
      Acta Cryst. A28 (1972) 629-635

(282) D.E. Williams
      Acta Cryst. A30 (1974) 71-77

(283) D.E. Williams and T.L. Starr
      Computers & Chem. 1 (1977) 173-177

(284) D.E. Williams
      Top. Curr. Phys. 26 (1981) 3-40

(285) F. Woldbye and Kj. Rasmussen
      Euroanalysis III (D.M. Carroll, Ed.), Appl. Sci. Publ., London
      (1979) p. 331-350

(286) P.E.S. Wormer and A. van der Avoird
      J. Chem. Phys. 62 (1975) 3326-3339

(287) A. Yokozeki and K. Kuchitsu
      Bull. Chem. Soc. Japan 44 (1971) 2926-2930

(288) V.B. Zhurkin, V.I. Poltev and V.L. Florentiev
      Mol. Biol. 14 (1980) 116-1130

Index of names

Subject index

A

ab initio, 35, 64, 87, 195
alcohols, 119
AMBER, 26
anharmonicity, 21
anomeric, 84
anomeric ratio, 50
anti, 8
asymmetric unit, 132
average structure, 11

B

bacterial endotoxins, 49
biological structure, 207
black box, 18, 157
Buckingham, 29

C

CFF, 157
charge, 81
charges, 71
CHARMM, 26, 157
chelate ring, 40
COLOR, 195
command list, 201
configuration, 6
conformational path, 51
constitution, 5
convergence
  acceleration, 131
  constant, 141
coordinate system, 132
coordination chemistry, 20
Coulomb, 21, 34, 142, 146
cross terms, 21, 25, 28
crown ether, 120
crystal properties, 88

D

data transport, 80
dielectric constant, 142, 146
dihedral angles, 44

D (continued)

dipole moment, 81
drug design, 208
dynamic stereochemistry, 28

E

ECEPP, 18, 33, 157
ECEPP/2, 33
Eckart conditions, 29
EDITOR, 165, 193
Einstein sum, 121
electron diffraction, 82
enantiomerisation, 28
endotoxic principle, 49
enthalpy, 8
EPEN, 18, 34, 157
EPEN/2, 34
equilibrium
  bond-length, 13
  structure, 82
ethers, 119
experimental data, 79

F

force constant, 13, 15
FORTRAN, 207
frequency, 81

G

gauche, 8
GAUSSIAN, 35, 63
GAUSSIAN-70, 22, 195
geminal, 20, 27-28
genetic alphabet, 207
GVFF, 24-25

H

Hessian, 74, 77, 124

Structure and Bonding

Editors: M.J.Clarke,
J.B.Goodenough, P.Hemmerich
J.A.Ibers, C.K.Jørgensen,
J.B.Neilands, D.Reinen,
R.Weiss, R.J.P.Williams

Volume 50

Topics in Inorganic and Physical Chemistry

1982. 48 figures, 29 tables. V, 178 pages. ISBN 3-540-11454-8

Contents: *J.A.Ibers, L.J.Pace, J.Martinsen, B.M.Hoffman:* Stacked Metal Complexes: Structures and Properties. – *M.J.Clarke, P.H.Fackler:* The Chemistry of Technetium: Toward Improved Diagnostic Agents. – *R.J.P.Williams:* The Chemistry of Lanthanide Ions in Solution and in Biological Systems. – *C.K.Jørgensen, R.Reisfeld:* Uranyl Photophysics.

Volume 51
A.Schweiger

Electron Nuclear Double Resonance of Transition Metal Complexes with Organic Ligands

1982. 47 figures, 19 tables. VIII, 128 pages. ISBN 3-540-11072-0

Contents: Introduction. – ENDOR-Instrumentation. – Analysis of ENDOR Spectra. – Advances ENDOR Techniques. – Interpretation of Hyperfine and Quadrupole Data. – Discussion of the Literature. – Concluding Remarks. – Appendix A: Abbreviations Used in this Paper. – Appendix B: Second Order ENDOR Frequencies. – Appendix C: Relations Between Nuclear Quadrupole Coupling Constants in Different Expressions of H_Q (Sect.5.2). – References. – Subject Index.

Springer-Verlag
Berlin
Heidelberg
New York
Tokyo

Volume 52

Structures versus Special Properties

1982. 90 figures, 19 tables. V, 202 pages. ISBN 3-540-11781-4

Contents: *R.G.Woolley:* Natural Optical Activity and the Molecular Hypothesis. – *L.Banci, A.Bencini, C.Benelli, D.Gatteschi, C.Zanchini:* Spectral-Structural Correlations in High-Spin Cobalt (II) Complexes. – *A.Tressaud, J.-M.Dance:* Relationships Between Structure and Low-Dimensional Magnetism in Fluorides. – *V.K.Jain, R.Bohra, R.C.Mehrotra:* Structure and Bonding in Organic Derivatives of Antimony (V).

Structure and Bonding

Editors: M.J.Clarke,
J.B.Goodenough, P.Hemmerich,
J.A.Ibers, C.K.Jørgenson,
J.B.Neilands, D.Reinen,
R.Weiss, R.J.P.Williams

Springer-Verlag
Berlin
Heidelberg
New York
Tokyo

Volume 53

Copper, Molybdenum, and Vanadium in Biological Systems

1983. 67 figures, 13 tables. V, 166 pages
ISBN 3-540-12042-4

Contents: *E.I.Solomon, K.W.Penfield, D.E.Wilcox:* Active Sites in Copper Proteins. An Electronic Structure Overview. – *B.A.Averill:* Fe-S and Mo-Fe-S Clusters as Models for the Active Site of Nitrogenase. – *N.D.Chasteen:* The Biochemistry of Vanadium. – *K.Kustin, G.C.McLeod, T.R.Gilbert, LeB.R.Briggs:* Vanadium and Other Metal Ions in the Physiological Ecology of Marine Organisms.

Volume 54

Inorganic Elements in Biochemistry

1983. 57 figures, 22 tables. V, 180 pages
ISBN 3-540-12542-6

Contents: *J.D.Odom:* Selenium Biochemistry – Chemical and Physical Studies. – *M.Lammers, H.Follmann:* The Ribonucleotide Reductases – A Unique Group of Metalloenzymes Essential for Cell Proliferation. – *P.H.Connett, K.E.Wetterhahn:* Metabolism of the Carcinogen Chromate by Cellular Constituents. – *S.Mann:* Mineralization in Biological Systems.

Volume 55

Transition Metal Complexes – Structures and Spectra

1984. 93 figures, 21 tables. V, 209 pages
ISBN 3-540-12833-6

Contents: *M.H.Gubelmann, A.F.Williams:* The Structure and Reactivity of Dioxygen Complexes of the Transition Metals. – *M.Bacci:* The Role of Vibronic Coupling in the Interpretation of Spectroscopic and Structural Properties of Biomolecules. – *F.Valach, B.Koreň, P.Sivý, M.Melník:* Crystal Structure Non-Rigidity of Central Atoms for Mn(II), Fe(II), Fe(III), Co(II), Co(III), Ni(II), Cu(II) and Zn(II) Complexes. – *F.Mathey, J.Fischer, J.J.Nelson:* Complexing Modes of the Phosphole Moiety.